JUSTICE AND HEALTH CARE

PHILOSOPHY AND MEDICINE

Editors:

H. TRISTRAM ENGELHARDT, JR.

Kennedy Institute of Ethics, Georgetown University, Washington, D.C., U.S.A.

STUART F. SPICKER

University of Connecticut Health Center, Farmington, Conn., U.S.A.

VOLUME 8

JUSTICE
AND HEALTH CARE

Edited by

EARL E. SHELP

Institute of Religion and Baylor College of Medicine, Houston, Texas, U.S.A.

D. REIDEL PUBLISHING COMPANY

DORDRECHT : HOLLAND / BOSTON : U.S.A.
LONDON : ENGLAND

174.2
J98
c.

Library of Congress Cataloging in Publication Data

Main entry under title:

Justice and health care.

(Philosophy and medicine ; v. 8)
Includes bibliographies and indexes.
1. Medical ethics.   2. Justice.   3. Medical policy.
I.   Shelp, Earl E., 1947–   [DNLM: 1.   Delivery of health care.
2.   Philosophy, Medical. W3 PH609 v. 8 / W 61 J96]
R724.J87   174'.2   80-26152
ISBN 90-277-1207-7
ISBN 90-277-1251-4 pbk (Pallas edition)

Published by D. Reidel Publishing Company,
P.O. Box 17, 3300 AA Dordrecht, Holland.

Sold and distributed in the U.S.A. and Canada
by Kluwer Boston Inc.,
190 Old Derby Street, Hingham, MA 02043, U.S.A.

In all other countries, sold and distributed
by Kluwer Academic Publishers Group,
P.O. Box 322, 3300 AH Dordrecht, Holland.

D. Reidel Publishing Company is a member of the Kluwer Group.

All Rights Reserved
Copyright © 1981 by D. Reidel Publishing Company, Dordrecht, Holland
and copyright holders as specified on appropriate pages within
No part of the material protected by this copyright notice may be reproduced or
utilized in any form or by any means, electronic or mechanical,
including photocopying, recording or by any information storage and
retrieval system, without written permission from the copyright owner

Printed in The Netherlands

TABLE OF CONTENTS

EARL E. SHELP / Introduction   vii

SECTION I / HISTORICAL AND
CONCEPTUAL BACKGROUND

ALLEN BUCHANAN / Justice: A Philosophical Review   3
MARTIN P. GOLDING / Justice and Rights: A Study in Relationship   23
FREDERICK S. CARNEY / Justice and Health Care: A Theological
    Review   37
LAURENCE B. McCULLOUGH / Justice and Health Care: Historical
    Perspectives and Precedents   51

SECTION II / ISSUES OF MICRO-ALLOCATION

ERIC J. CASSELL / Do Justice, Love Mercy: The Inappropriateness
    of the Concept of Justice Applied to Bedside Decisions   75
MARC LAPPÉ / Justice and Prenatal Life   83
ALBERT R. JONSEN / Justice and the Defective Newborn   95
MICHAEL D. BAYLES / Justice and the Dying   109

SECTION III / ISSUES OF MACRO-ALLOCATION

H. TRISTRAM ENGELHARDT, JR. / Health Care Allocations:
    Responses to the Unjust, the Unfortunate, and the Undesirable   121
JAMES F. CHILDRESS / Priorities in the Allocation of Health Care
    Resources   139
BARUCH BRODY / Health Care for the Haves and Have-nots:
    Toward a Just Basis of Distribution   151
KIM CARNEY / Cost Containment and Justice   161
KAREN LEBACQZ / Justice and Human Research   179

RONALD M. GREEN / Justice and the Claims of Future Generations 193
EARL E. SHELP / Justice: A Moral Test for Health Care and Health Policy 213

NOTES ON CONTRIBUTORS 231

INDEX 233

EARL E. SHELP

INTRODUCTION

Bioethics is a discipline still not fully explored in spite of its rather remarkable expansion and sophistication during the past two decades. The proliferation of courses in bioethics at educational institutions of every description gives testimony to an intense academic interest in its concerns. The media have catapulted the dilemmas of bioethics out of the laboratory and library into public view and discussion with a steady report of the so-called 'miracles of modern medicine' and the moral perplexities which frequently accompany them. The published work of philosophers, theologians, lawyers and others represents a substantial and growing body of literature which explores relevant concepts and issues. Commitments have been made by existing institutions, and new institutions have been chartered to further the discussion of the strategic moral concerns that attend recent scientific and medical progress.

This volume focuses attention on one of the numerous topics of interest within bioethics. Specifically, an examination is made of the implications of the principle of justice for health care. Apart from four essays in *Ethics and Health Policy* edited by Robert Veatch and Roy Branson [4] the discussion of justice and health care has been occasional, almost non-existent, and scattered. The paucity of literature in this area is regrettable but perhaps understandable. On the one hand, Joseph Fletcher, one of the contemporary pioneers in bioethics, can hold that "distributive justice is the core or key question for biomedical ethics" ([1], p. 102). On the other hand, Paul Ramsey can characterize the task of rationally ordering medical priorities and overall social priorities as "the most incorrigible social and ethical question" ([3], p. 268). An appreciation of the fundamental importance of justice to the concerns of bioethics will not permit it to be forgotten. Yet, a recognition of the profound difficulty of its explication probably accounts for the scarcity of published efforts to do so.

This collection of essays advances the discussion by addressing selected matters of historical and conceptual interest. No claims to completeness or finality are made for the volume as a whole or for any one of its parts. On the contrary, a sense of the work that remains to be done is implicit within each part. As a result, much of the volume can be characterized as

Earl E. Shelp (ed.), Justice and Health Care, vii–xiv.
Copyright © 1981 by D. Reidel Publishing Company.

an exploration into previously uncharted terrain, or as an attempt to clarify and analyze ideas. To these ends, the essays are grouped into three broad sections of investigation. The first section contains four essays providing background conceptual and historical analyses. The second section consists of four essays under the rubric of 'Issues of Micro-Allocation'. Here the interest is to explore how considerations of justice influence the care of classes of individuals, and the relationships between providers and recipients of health care services. The third, 'Issues of Macro-Allocation', is the most extensive. Seven contributors focus on the broader questions of allocation and the role of justice in health policy decisions.

The section of background essays opens with Allen Buchanan's review of three major theories of justice — utilitarianism, Rawls's 'Justice as Fairness', and libertarianism. He considers the implications of these theories for moral problems in health care by confronting them with four questions which focus on a right to health care, the relative priority and relationship of health care to other goods, the relationship of various forms of health care to each other, and the justice or injustice of the current health care system. He concludes that the examined theories can offer, at best, only ambiguous answers to what he considers the most difficult and pressing issues with which a theory of justice must cope.

Martin Golding's paper engages in a search for the meaning of 'right' and its connection to 'justice'. He traces the development of these notions as they were articulated by the classicists and subsequently interpreted by commentators into the early modern era. He suggests that 'right' was understood by the classicists in an 'objective sense' of what is right. The 'subjective sense' of something one has, according to Golding, did not materialize as a distinctive idea until the Middle Ages. Even then rights were seen as derived from justice or 'the right' and correspond to what is termed 'claim-rights' or an entitlement to a good. Golding claims that the notion of 'liberty-rights' may have its origin in the 16th century but becomes dominant in the 17th and 18th centuries. He concludes that on the basis of his historical review that rights derive from justice and that a merit notion of rights (one form of a welfare-right conception) is the more fundamental idea.

A review of theological perspectives is then provided. Jewish, Christian, and Islamic understandings of justice are explored in order to indicate how religious understandings frame notions of justice. Fred Carney in this essay indicates the religious roots of the notion that the poor make a claim of distributive justice, not just of charity, upon the believer. He traces as well the interplay of secular religious concepts of justice with sentiments drawn from religious roots. In addition, he indicates the special difficulty for religious views in reconciling the great misfortune of many in sickness and debility with the notion of a just God. It is, after all, within a theological perspective that the line between the unfair and the unjust, a point raised later in this volume by Engelhardt, may be most difficult to maintain.

Larry McCullough reminds us that recent inquiries into 'institutional medical ethics' are not the first efforts to consider justice and health care. He suggests that recent calls for reform of the present system rest on one of four foundations: a right to health care, utilitarian theory, retributive and distributive theories of justice, and appeals to virtue and obligation. These efforts to articulate and ground an 'institutional medical ethic' have precedents in the 18th and 19th centuries. He cites as evidence of his claim the debate in France during their Revolution about the natural right to health care, the German concept of medical police grounded in a view that duties are owed to the state, and the British and North American concept that the duties of a virtuous physician owed to the patient ought to be extended to the public. He concludes with (1) a demonstration of how these precedents can inform the current discussions of justice and health care (an example of institutional medical ethics), and (2) the caution that present solutions to the complex and difficult contemporary problems of balancing interests may be no more permanent than those that have come before.

The focus shifts from historical inquiry to issues of more personal concern in the essay of Eric Cassell. Cassell opens the second section with the argument that "it is usually not appropriate to ground individual treatment decisions on ideas of justice." He grants that while justice may be an appropriate consideration at the level of public policy it has no place at the bedside. He urges that the concepts of love of humankind, compassion, and mercy are the better guides for moral conduct at this level. He perceives an emphasis on justice at the bedside as a threat to good medicine which must recognize and respond to individual differences and inequalities. He laments the contemporary failure to embrace the virtues of love, compassion, and mercy as "the moral basis of behavior toward the sick." He warns that justice without a compassion which recognizes fundamental differences between persons can be oppressive. Cassell concludes that wisdom, not justice, must guide compassion and mercy, as difficult decisions are faced in the care of the sick.

Marc Lappé explains that a consideration of justice for prenatal life is only now becoming possible because of technological advances that enable intervention and a social appreciation of fetal life as worthy of protection.

He draws upon his experience in the California Department of Health Services to illustrate how prenatal care initiatives can be informed by considerations of distributive, retributive, and social justice. He suggests that the Supreme Court assessment in *Roe v. Wade* of fetal status accruing incrementally as the potential for separate existence increases is an inadequate basis for public policy. He endorses a biological model as more appropriate since it requires a recognition of the importance of genetic and environmental events to 'normal' fetal development and 'normal' outcomes of pregnancy which are the desired ends of justice for prenatal life. He holds that justice for prenatal life requires an equal opportunity for minimal well-being. The focus of distributive justice is on access to fertility and prenatal services. The focus of retributive justice is on the protection of vulnerable prenatal life and compensation for certain sustained injuries. And the focus of social justice is on redressing the relative disadvantage of the least well-off group in society with regard to fetal outcomes.

The requirements of justice in the care of a class of patients termed defective newborns are considered by Albert Jonsen. He defends 'quality of life' considerations for treatment which he defined in an earlier work [2] and iterates here. He agrees that only medical therapy which, at its initiation, "promises the establishment of a basic equality among all newborns" is required. He understands basic equality to consist in "possessing those integrated functions of major physical and physiological systems (including cerebral capacity for cognitive life) necessary for sustaining life." When this prospect is not realistic it is not unjust to withhold treatment. He concludes his discussion with a comment on the question of the cost-effectiveness of neonatal intensive care. He feels that the ethics of this sort of policy turns on two considerations: (1) "the fairness of quality criteria," and (2) "empirical data ... about the costs and benefits of making decisions in accord with such a position."

The final essay in the second section is by Michael Bayles. He suggests that the appropriateness of justice to a discussion of the care of the dying depends on one's concept of justice. He discusses three options: (1) justice as respecting rights, (2) distributive justice and a concern with possible conflicts with efficiency, and (3) compensatory justice and the question of the entitlement of the dying to benefits in compensation of their state. He basically addresses the question of whether or not a terminal condition is a relevant difference for a distribution of care and resources. He concludes that dying, as he defines it, may be a relevant reason to deny care because the expected benefits will not ensue. It may also be a relevant reason to provide extra care provided that empirical evidence can support the claim that the target condition is correlated to the fact of dying. Yet, he is uneasy with an appeal to justice in the care of the dying because it invokes an image of a relationship between strangers. He prefers to view health care in general, and care of the dying in particular, as a personal, familiar, and responsive encounter. To base decisions in this area on considerations of justice risks, in his view, depersonalizing and publicizing a very personal and private experience.

The final section of essays begins with a contribution by Tristram Engelhardt. Engelhardt puzzles over the different ways in which health care distributions can be viewed with respect to ethical obligations. He identifies three genre of systems of distribution (pure free market, mixed system, and egalitarian) and attempts to discover the significance of ethical claims for health care allocations. He states that such endeavors must cope with several confusions. What is the priority of claims to health care when compared with other claims? What is the priority of various kinds of claims to health care? How can one view the natural lottery? How can one view the ownership of goods? How these confusions are classified will influence one's view of the morality of the three genre of systems of health care distribution. Engelhardt offers an alternative model to help sort out intuitions and arguments regarding health care allocations which admits an absence of absolute answers. He suggests that notions of a right to health care are created, as much as discovered, based upon a culture, its history, and its values.

James Childress continues the discussion of priorities and offers an analysis of some of the ethical issues in the allocation of health care. He focuses on two questions: (1) How much resources should be in health care versus other social goods, and (2) once an amount for health care is set, how much should be invested in prevention and how much for rescue or crisis medicine. He states that the answer to the first question depends on one's definition of health, its value, its causes, and whether there is a right to a decent minimum of health care, and what that minimum is. He admits that the current emphasis of allocation within the health care budget favors crisis care. Yet he feels that it is difficult to know what mix of preventive and crisis care will eventuate in reduced morbidity and premature mortality. If the emphasis shifts to prevention, he indicates that four considerations (symbolic value of rescue efforts, principle of fairness, principle of equal access, principle of liberty) will show it not to be a simple matter.

Baruch Brody considers the problem of providing care to the medically indigent (have-nots) at a cost affordable to taxpayers (haves). He argues that the focus on health care distribution has been too narrow. Instead, he urges a broader view which sees health care "in the context of programs of redistribution aimed at promoting greater equality." Brody sketches a new theory of justice called 'quasi-libertarianism' which has implications for health care distribution. He challenges the traditional libertarian notion of property right. He feels that the traditional notion does not justify an entitlement to non-labor-created wealth, e.g., natural resources. He proposes that these resources should be leased with the proceeds paid into a social insurance fund. Cash disbursements would be made to eligible recipients in order to assist the pursuit of their intrinsic ends. Such a theory, according to Brody, would lead to a redistribution of wealth and not to particular rights to any one basic need including health care. This system of redistribution would eliminate the macro-problem of justice in providing health care to the have-nots.

Kim Carney reflects on current efforts to contain the costs of health care. She explores the implications of these efforts for a just health care system. These containment proposals have been in the form of regulation or stimulation of competition. She feels that these reviewed endeavors do not present insurmountable questions of justice, since no particular single group would bear an excessive burden. She is concerned, however, with the ability of the less articulate members of society to advocate their interest. Carney is hopeful, however, that these dangers can be avoided with proper monitoring.

Karen Lebacqz's essay considers the role of justice in human research. She sees the concern of justice (i.e., selection) in human research as an important addition to the principles of respect for persons and beneficence. Lebacqz argues the position that justice "requires an equal distribution of benefits and burdens except where unequal distribution is justified by consideration of merit, social utility, or prior harm." She demonstrates how this principle can be applied to the selection of subjects and compensation for injury. She concludes that informed consent and a reasonable risk-benefit ratio is no longer sufficient to establish the ethical acceptability of human research. Considerations of justice must now be included.

Ronald Green's contribution addresses the question of health-related responsibilities to future generations. He urges a mediating view between the extreme answers of everything and nothing. He suggests that the social contract approach of Rawls is profitable to a reasoned response to the question. He thinks that obligations exist to the future because it is "proper that a moral relationship exist among those separated in time." He adopts the general principle that "we are required to strive to ensure that our

INTRODUCTION   xiii

descendents are left with the means to a progressively better quality of life than ourselves, and, at a very minimum, are not rendered worse off by our actions." The implications of this general principle for health care are twofold. First, efforts to expand access to good health care services are supported. Second, certain existing or projected health care priorities are questionable when their effect is extrapolated into the future. A long-term view would lead to, according to Green, a current emphasis on environmental medicine, biomedical research, and attending the health care needs of less-advantaged groups.

Earl Shelp's essay concludes the collection. Shelp reflects on the difficulty of arriving at a conception of justice and its relevance to health care distributions that all would accept. He finds points of agreement within four theories of justice but argues that a strong concensus for an application to health care is problematic because the theories are value-laden. He understands concepts of justice as fluid and responsive to changed social values and changed material circumstance. He claims, however, that health care and health policy are proper contemporary subjects of justice since each institution is associated with values and the means instrumental to human welfare. He concludes that justice is a legitimate, necessary moral test for health care and health policy but that it is not a sufficient test to establish the morality of health care and health policy.

Collections of this sort offer certain advantages and disadvantages. Positively they provide a variety of perspectives on a critical issue by drawing upon the distinctive expertise of a number of scholars. Negatively they fail to provide a complete and systematic treatment of an important issue. If they achieve their purpose, they encourage further discussions rather than discourage them. They also clarify some issues but pose others perhaps equally perplexing. The moral problem of justice and health care merits our most thoughtful consideration. The importance of justice to the moral character of American society and one of its most valued institutions cannot be almost ignored any longer. It is hoped that this collection will occasion a sustained, informative, and substantive explication of its two concerns — justice and health care.

This project would not have been possible without the enthusiastic interest and commitment of each of the contributors. My debt to each is gratefully acknowledged. The patience, encouragement, and counsel of the series editors, H. Tristram Engelhardt, Jr., and Stuart F. Spicker, have been essential to the completion of this collection. Their able assistance, along with that of Susan M. Engelhardt, has enhanced the quality of this work and is deeply appreciated. Gratitude is also expressed to Mrs. Audrey Laymance whose energies and typing skills are unmatched. My wish is that each one has benefited from his participation in this work at least to a degree equal to mine.

March, 1980

BIBLIOGRAPHY

1. Fletcher, J.: 1976, 'Ethics and Health Care Delivery: Computers and Distributive Justice', in R. M. Veatch and R. Branson (eds.), *Ethics and Health Policy,* Ballinger Publishing Co., Cambridge, pp. 99–109.
2. Jonsen, A. R. and Garland, M. J. (eds.): 1976, *Ethics of Newborn Intensive Care,* University of California, Berkeley.
3. Ramsey, P.: 1970, *The Patient as Person,* Yale University Press, New Haven.
4. Veatch, R. M., and Branson, R. (eds.): 1976, *Ethics and Health Policy,* Ballinger Publishing Co., Cambridge.

SECTION I

HISTORICAL AND CONCEPTUAL BACKGROUND

ALLEN BUCHANAN

JUSTICE: A PHILOSOPHICAL REVIEW

I. INTRODUCTION

The past decade has seen the bourgeoning of bioethics and the resurgence of theorizing about justice. Yet until now these two developments have not been as mutually enriching as one might have hoped. Bioethicists have tended to concentrate on micro issues (moral problems of individual or small group decision making), ignoring fundamental moral questions about the macro structure within which the micro issues arise. Theorists of justice have advanced very general principles but have typically neglected to show how they can illuminate the particular problems we face in health care and other urgent areas.

Micro problems do not exist in an institutional vacuum. The parents of a severely impaired newborn and the attending neonatologist are faced with the decision of whether to treat the infant aggressively or to allow it to die because neonatal intensive care units now exist which make it possible to preserve the lives of infants who previously would have died. Neonatal intensive care units exist because certain policy decisions have been made which allocated certain social resources to the development of technology for sustaining defective newborns rather than for preventing birth defects. Limiting moral inquiry to the micro issues supports an unreasoned conservatism by failing to examine the health care institutions within which micro problems arise and by not investigating the larger array of institutions of which the health care sector is only one part. Since not only particular actions but also policies and institutions may be just or unjust, serious theorizing about justice forces us to expand the narrow focus of the micro approach by raising fundamental queries about the background social, economic, and political institutions from which micro problems emerge.

On the other hand, the attention to individual cases which dominates contemporary bioethics can provide a much needed concrete focus for refining and assessing competing theories of justice. The adequacy or inadequacy of a moral theory cannot be determined by inspecting the principles which constitute it. Instead, rational assessment requires an on-going process in which general principles are revised and refined through confrontation with the rich complexity of our considered judgments about particular cases, while our judgments about particular cases are gradually structured and modified by our provisional acceptance of general principles. Since our considered judgments about particular cases may often be more sensitive and sure than our assessments of abstract principles, careful attention to accurately described, concrete moral situations is essential for theorizing about justice.

Further, it is not just that the problems of bioethics provide one class of test cases for theories of justice among others: the problems of bioethics are among the most difficult and pressing issues with which a theory of justice must cope. It appears, then, that the continued development of both bioethics and of theorizing about justice in general requires us to explore the problems of justice in health care. In this essay I hope to contribute to that enterprise by first providing a sketch of three major theories of justice and by then attempting to ascertain some of their implications for moral problems in health care.

II. THEORIES OF JUSTICE

Utilitarianism

Utilitarianism purports to be a comprehensive moral theory, of which a utilitarian theory of justice is only one part. There are two main types of comprehensive utilitarian theory: Act and Rule Utilitarianism. Act Utilitarianism defines rightness with respect to particular acts: an act is right if and only if it maximizes utility. Rule Utilitarianism defines rights with respect to rules of action and makes the rightness of particular acts depend upon the rules under which those acts fall. A rule is right if and only if general compliance with that rule (or with a set of rules of which it is an element) maximizes utility, and a particular action is right if and only if it falls under such a rule.

Both Act and Rule Utilitarianism may be versions of either Classic or Average Utilitarianism. Classic Utilitarianism defines the rightness of acts or rules as maximization of *aggregate* utility; Average Utilitarianism defines rightness as maximization of utility *per capita*. The aggregate utility produced by an act or by general compliance with a rule is the sum of the utility produced for each individual affected. Average utility is the aggregate utility divided by the number of individuals affected. 'Utility' is defined as pleasure, satisfaction, happiness, or as the realization of preferences, as the latter are revealed through individuals' choices.

The distinction beween Act and Rule Utilitarianism is important for a utilitarian theory of justice, since the latter must include an account of when *institutions* are just. Thus, institutional rules may maximize utility even though those rules do not direct individuals as individuals or as occupants of institutional positions to maximize utility in a case by case fashion. For example, it may be that a judicial system which maximizes utility will do so by including rules which prohibit judges from deciding a case according to their estimates of what would maximize utility in that particular case. Thus the utilitarian justification of a particular action or decision may not be that *it* maximizes utility, but rather that it falls under some rule of an institution or set of institutions which maximizes utility.[1]

Some utilitarians, such as John Stuart Mill, hold that principles of justice are the most basic moral principles because the utility of adherence to them is especially great. According to this view, utilitarian principles of justice are those utilitarian moral principles which are of such importance that they may be *enforced*, if necessary. Some utilitarians, including Mill perhaps, also hold that among the utilitarian principles of justice are principles specifying individual *rights*, where the latter are thought of as enforceable claims which take precedence over appeals to what would maximize utility in the particular case. Indeed, some contemporary rights theorists such as Ronald Dworkin define a (justified) right claim as one which takes precedence over mere appeals to what would maximize utility.

A utilitarian moral theory, then, can include rights principles which themselves prohibit appeals to utility maximization, so long as the justification of those principles is that they are part of an institutional system which maximizes utility. In cases where two or more rights principles conflict, considerations of utility may be invoked to determine which rights principles are to be given priority. Utilitarianism is incompatible with rights only if rights exclude appeals to utility maximization at all levels of justification, including the most basic institutional level. Rights founded ultimately on considerations of utility may be called *derivative*, to distinguish them from rights in the *strict* sense.

Utilitarianism is the most influential version of *teleological* moral theory. A moral theory is teleological if and only if it defines the good independently of the right and defines the right as that which maximizes the good. Utilitarianism defines the good as happiness (satisfaction, etc.), independently of any account of what is morally right, and then defines the right as that which maximizes the good (either in the particular case or at the institutional level). A moral theory is *deontological* if and only if it is not a teleological theory, i.e., if and only if it either does not define the good independently of the right or does not define the right as that which maximizes the good. Both the second and third theories of justice we shall consider are deontological theories.

John Rawls's Theory: Justice as Fairness
In *A Theory of Justice* Rawls pursues two main goals. The first is to set out a small but powerful set of principles of justice which underlie and explain the considered moral judgments we make about particular actions, policies, laws, and institutions. The second is to offer a theory of justice superior to Utilitarianism. These two goals are intimately related for Rawls because he believes that the theory which does a better job of supporting and accounting for our considered judgments is the better theory, other things being equal. The principles of justice Rawls offers are as follows:

(1) The principle of greatest equal liberty:

Each person is to have an equal right to the most extensive system of equal basic liberties compatible with a similar system of liberty for all ([6], pp. 60, 201–205).

(2) The principle of equality of fair opportunity:

Offices and positions are to be open to all under conditions of equality of fair opportunity — persons with similar abilities and skills are to have equal access to offices and positions ([6], pp. 60, 73, 83–89).[2]

(3) The difference principle:

Social and economic institutions are to be arranged so as to benefit maximally the worst off ([6], pp. 60, 75–83).[3]

The basic liberties referred to in (1) include freedom of speech, freedom of conscience, freedom from arbitrary arrest, the right to hold personal property, and freedom of political participation (the right to vote, to run for office, etc.).

Since the demands of these principles may conflict, some way of ordering them is needed. According to Rawls, (1) is *lexically prior* to (2) and (2) is *lexically prior* to (3). A principle 'P' is lexically prior to a principle 'Q' if and only if we are first to satisfy all the requirements of 'P' before going on to satisfy the requirements of 'Q'. Lexical priority allows no trade-offs between the demands of conflicting principles: the lexically prior principle takes absolute priority.

Rawls notes that "many kinds of things are said to be just or unjust: not only laws, institutions, and social systems, but also particular actions . . . decisions, judgments and imputations . . ." ([6], p. 7). But he insists that the primary subject of justice is the *basic structure* of society because it exerts a pervasive and profound influence on individuals' life prospects. The basic structure is the entire set of major political, legal, economic, and social institutions. In our society the basic structure includes the Constitution, private ownership of the means of production, competitive markets, and the monogamous family. The basic structure plays a large role in distributing the burdens and benefits of cooperation among members of society.

If the primary subject of justice is the basic structure, then the primary problem of justice is to formulate and justify a set of principles which a just basic structure must satisfy. These principles will specify how the basic structure is to distribute prospects of what Rawls calls *primary goods*. These include the basic liberties (listed above under (1)), 'as well as powers, authority, opportunities, income, and wealth. Rawls says that primary goods are things that every rational person is presumed to want, because they normally have a use, whatever a person's rational plan of life ([6], p. 62). Principle (1) regulates the distribution of prospects of basic liberties; (2) regulates the distribution of prospects of powers and authority, so far as these are attached to institutional offices and positions, and (3) regulates the distribution of prospects of the other primary goods, including wealth and income. Though the first and second principles require equality, the difference principle allows inequalities so long as the total system of institutions of which they are a part maximizes the prospects of the worst off to the primary goods in question.

Rawls advances three distinct types of justification for his principles of justice. Two appeal to our considered judgments, while the third is based on what he calls the Kantian interpretation of his theory.

The first type of justification rests on the idea, mentioned earlier, that if a set of principles provides the best account of our considered judgments about what is just or unjust, then that is a reason for accepting those principles. A set of principles accounts for our judgments only if those judgments can be derived from the principles, granted the relevant facts for their application.

Rawls's second type of justification maintains that if a set of principles would be chosen under conditions which, according to our considered judgments, are appropriate conditions for choosing principles of justice, then this is a reason for accepting those principles. The second type of justification includes three parts: (1) A set of conditions for choosing principles of justice must be specified. Rawls labels the complete set of conditions the 'original position.' (2) It must be shown that the conditions specified are (according to our considered judgments) the appropriate conditions of choice. (3) It must be shown that Rawls's principles are indeed the principles which would be chosen under those conditions.

Rawls construes the choice of principles of justice as an ideal social contract. "The principles of justice for the basic structure of society are the principles that free and rational persons ... would accept in an initial situation of equality as defining the fundamental terms of their association" ([6], p. 11). The idea of a social contract has several advantages. First, it allows us to view principles of justice as the object of a *rational collective choice*. Second, the idea of *contractual obligation* is used to emphasize that the choice expresses a basic commitment and that the principles agreed on may be rightly enforced. Third, the idea of a contract as a *voluntary agreement* which sets terms for mutual advantage suggests that the principles of justice should be "such as to draw forth the willing cooperation" ([6], p. 15) of all members of society, including those who are worse off.

The most important elements of the original position for our purposes are (a) the characterization of the parties to the contract as individuals who desire to pursue their own life plans effectively and who "have a highest-order interest in how ... their interests ... are shaped and regulated by social institutions" ([8], p. 641); (b) the 'veil of ignorance', which is a constraint on the information the parties are able to utilize in choosing principles of justice; and (c) the requirement that the principles are to be chosen on the assumption that they will be complied with by all (the universalizability condition) ([6], p. 132).

The parties are characterized as desiring to maximize their shares of primary goods, because these goods enable one to implement effectively the widest range of life plans and because at least some of them, such as freedom of speech and of conscience, facilitate one's freedom to choose and revise one's life plan or conception of the good. The parties are to choose "from behind a veil of ignorance" so that information about their own particular characteristics or social positions will not lead to bias in the choice of principles. Thus they are described as not knowing their race, sex, socioeconomic, or political status, or even the nature of their particular conceptions of the good. The informational restriction also helps to insure that the principles chosen will not place avoidable restrictions on the individual's freedom to choose and revise his or her life plan.[4]

Though Rawls offers several arguments to show that his principles would be chosen in the original position, the most striking is the *maximin argument*. According to this argument, the rational strategy in the original position is to choose that set of principles whose implementation will maximize the minimum share of primary goods which one can receive as a member of society, and principles (1), (2), and (3) will insure the greatest minimal share. Rawls's claim is that because these principles protect one's basic liberties and opportunities and insure an adequate minimum of goods such as wealth and income (even if one should turn out to be among the worst off) the rational thing is to choose them, rather than to gamble with one's life prospects by opting for alternative principles. In particular, Rawls contends that it would be irrational to reject his principles and allow one's life prospects to be determined by what would maximize utility, since utility maximization might allow severe deprivation or even slavery for some, so long as this contributed sufficiently to the welfare of others.

Rawls raises an important question about this second mode of justification when he notes that this original position is purely hypothetical. Granted that the agreement is never actually entered into, why should we regard the principles as binding? The answer, according to Rawls, is that we do in fact accept the conditions embodied in the original position ([6], p. 21). The following qualification, which Rawls adds immediately after claiming that the conditions which constitute the original position are appropriate for the choice of principles of justice according to our considered judgments, introduces his third type of justification: "Or if we do not [accept the conditions of the original position as appropriate for choosing principles of justice] then *perhaps we can be persuaded to do so by philosophical reflections*" (emphasis added, [6], p. 21). In the Kantian interpretation section of *A Theory of Justice* Rawls sketches a certain kind of philosophical justification for the conditions which make up the original position (based on Kant's conception of the 'noumenal self' or autonomous agent).

For Kant an autonomous agent's will is determined by rational principles and rational principles are those which can serve as principles for all rational beings, not just for this or that agent, depending upon whether or not he has some particular desire which other rational beings may not have. Rawls invites us to think of the original position as the perspective from which autonomous agents see the world. The original position provides a "procedural interpretation" of Kant's idea of a Realm of Ends or community of "free and equal rational beings". We express our nature as autonomous agents when we act from principles that would be chosen in conditions which reflect that nature ([6], p. 252).

Rawls concludes that, when persons such as you and I accept those principles that would be chosen in the original position, we express our nature as autonomous agents, i.e., we act autonomously. There are three main grounds for this thesis, corresponding to the three features of the original position cited earlier. First, since the veil of ignorance excludes information about any particular desires which a rational agent may or may not have, the choice of principles is not determined by any particular desire. Second, since the parties strive to maximize their share of primary goods, and since primary goods are attractive to them because they facilitate freedom in choosing and revising life plans and because they are flexible means not tied to any particular ends, this is another respect in which their choice is not determined by particular desires. Third, the original position includes the requirement that the principles of justice must be universalizable and this is to insure that they will be principles for rational agents in general and not just for agents who happen to have this or that particular desire.

In the *Foundations of the Metaphysics of Morals* Kant advances a moral philosophy which identifies autonomy with rationality [4]. Hence for Kant the question "Why should one express our nature as autonomous agents?" is answered by the thesis that rationality requires it. Thus *if* Rawls's third type of justification succeeds in showing that we best express our autonomy when we accept those principles in the belief that they would be chosen from the original position, and *if* Kant's identification of autonomy with rationality is successful, the result will be a justification of Rawls's principles which is distinct from both the first and second modes of justification. So far as this third type of justification does not make the acceptance of Rawls's principles hinge on whether the principles themselves or the conditions from which they would be chosen match our considered judgments, it is not directly vulnerable either to the charge that Rawls has misconstrued our considered judgments or that congruence with considered judgments, like the appeal to mere consensus, has no justificatory force.

It is important to see that Rawls understands his principles of justice as principles which generate *rights* in what I have called the strict sense. Claims based upon the three principles are to take precedence over considerations of utility and the principles themselves are not justified on the grounds that a basic structure which satisfies them will maximize utility. Moreover, Rawls's theory is not a teleological theory of any kind because it does not define the right as that which maximizes the good, where the good is defined independently of the right. Instead it is perhaps the most influential current instance of a deontological theory.

Nozick's Libertarian Theory

There are many versions of libertarian theory, but their characteristic doctrine is that coercion may only be used to prevent or punish physical harm, theft, and fraud, and to enforce contracts. Perhaps the most influential and systematic recent instance of Libertarianism is the theory presented by Robert Nozick in *Anarchy, State, and Utopia* [5]. In Nozick's theory of justice, as in libertarian theories generally, the right to private property is fundamental and determines both the legitimate role of the state and the most basic principles of individual conduct.

Nozick contends that individuals have a property right in their persons and in whatever 'holdings' they come to have through actions which conform to (1) "the principle of justice in [initial] acquisition" and (2) "the principle of justice in transfer" ([5], p. 151). The first principle specifies the ways in which an individual may come to own hitherto unowned things without violating anyone else's rights. Here Nozick largely follows John Locke's famous account of how one makes natural objects one's own by "mixing one's labor" with them or improving them through one's labor. Though Nozick does not actually formulate a principle of justice in (initial) acquisition, he does argue that whatever the appropriate formulation is it must include a 'Lockean Proviso', which places a constraint on the holdings which one may acquire through one's labor. Nozick maintains that one may appropriate as much of an unowned item as one desires so long as (a) one's appropriation does not worsen the conditions of others in a special way, namely, by creating a situation in which others are "no longer ... able to use freely [without exclusively appropriating] what [they] ... previously could" or (b) one properly compensates those whose condition is worsened by one's appropriation in the way specified in (a) ([5], pp. 178–179). Nozick emphasizes that the Proviso only picks out one way in which one's appropriation may worsen the condition of others; it does not forbid appropriation or require compensation in cases in which one's appropriation of an unowned thing worsens another's condition merely by limiting his opportunities to appropriate (rather than merely use) that thing, i.e., to make it his property.

The second principle states that one may justly transfer one's legitimate holdings to another through sale, trade, gift, or bequest and that one is entitled to whatever one receives in any of these ways, so long as the person from whom one receives it was entitled to that which he transferred to you. The right to property which Nozick advances is the right to exclusive control over anything one can get through initial appropriation (subject to the Lockean Proviso) or through voluntary exchanges with others entitled to what they transfer. Nozick concludes that a distribution is just if and only if it arose from another just distribution by legitimate means. The principle of justice in initial acquisition specifies the legitimate 'first moves,' while the principle of justice in transfers specifies the legitimate ways of moving from one distribution to another: "Whatever arises from a just situation by just steps is itself just" ([5], p. 151).

Since not all existing holdings arose through the 'just steps' specified by the principles of justice in acquisition and transfer, there will be a need for a *principle of rectification* of past injustices. Though Nozick does not attempt to formulate such a principle he thinks that it might well require significant redistribution of holdings.

Apart from the case of rectifying past violations of the principles of acquisition and transfer, however, Nozick's theory is strikingly anti-redistributive. Nozick contends that attempts to force anyone to contribute any part of his legitimate holdings to the welfare of others is a violation of that person's property rights, whether it is undertaken by private individuals or the state. On this view, coercively backed taxation to raise funds for welfare programs of any kind is literally theft. Thus, a large proportion of the activities now engaged in by the government involve gross injustices.

After stating his theory of rights, Nozick tries to show that the state is legitimate so long as it limits its activities to the enforcement of these rights and eschews redistributive functions. To do this he employs an 'invisible hand explanation,' which purports to show how the minimal state could arise as an unintended consequence of a series of voluntary transactions which violate no one's rights. The phrase 'invisible hand explanation' is chosen to stress that the process by which the minimal state could emerge fits Adam Smith's famous account of how individuals freely pursuing their own private ends in the market collectively produce benefits which are not the aim of anyone.

The process by which the minimal state could arise without violating anyone's rights is said to include four main steps ([5], pp. 10–25).[5] First, individuals in a 'state of nature' in which (libertarian) moral principles are generally respected would form a plurality of 'protective agencies' to enforce their libertarian rights, since individual efforts at enforcement would be inefficient and liable to abuse. Second, through competition for clients, a 'dominant protective agency' would eventually emerge in a given geographical area. Third, such an agency would eventually become a 'minimal state' by asserting a claim of monopoly over protective services in order to prevent less reliable efforts at enforcement which might endanger its clients: it would forbid 'independents' (those who refused to purchase its services) from seeking other forms of enforcement. Fourth, again assuming that correct moral principles are generally followed, those belonging to the dominant protective agency would compensate the 'independents', presumably by providing them with free or partially subsidized protection services. With the exception of taxing its clients to provide compensation for the independents, the minimal state would act only to protect persons against physical injury, theft, fraud, and violations of contracts.

It is striking that Nozick does not attempt to provide any systematic *justification* for the Lockean rights principles he advocates. In this respect he departs radically from Rawls. Instead, Nozick assumes the correctness of the Lockean principles and then, on the basis of that assumption, argues that the minimal state and only the minimal state is compatible with the rights those principles specify.

He does, however, offer some arguments against the more-than-minimal state which purport to be independent of that particular theory of property rights which he assumes. These arguments may provide indirect support for his principles insofar as they are designed to make alternative principles, such as Rawls's, unattractive. Perhaps most important of these is an argument designed to show that any principle of justice which demands a certain distributive end state or pattern of holdings will require frequent and gross disruptions of individuals' holdings for the sake of maintaining that end state or pattern. Nozick supports this general conclusion by a vivid example. He asks us to suppose that there is some distribution of holdings 'D_1' which is required by some end-state or patterned theory of justice and that 'D_1' is achieved at time 'T'. Now suppose that Wilt Chamberlain, the renowned basketball player, signs a contract stipulating that he is to receive twenty-five cents from the price of each ticket to the home games in which he performs, and suppose that he nets $250,000 from this arrangement. We now have a new distribution 'D_2'. Is 'D_2' unjust? Notice that by hypothesis those who paid the price of admission were entitled to control over the resources they held in 'D_1', (as were Chamberlain and the team's owners). The new distribution arose through *voluntary exchanges of legitimate holdings*, so it is difficult to see how it could be unjust, even if it does diverge from 'D_1'. From this and like examples, Nozick concludes that attempts to maintain any end-state or patterned distributive principle would require continuous interference in peoples' lives ([5], pp. 161–163).

As in the cases of Utilitarianism and Rawls's theory, Nozick and libertarians generally do not limit morality to justice. Thus, Nozick and others emphasize that a libertarian theory of individual rights is to be supplemented by a libertarian theory of virtues which recognizes that not all moral principles are suitable objects of enforcement and that moral life includes more than the nonviolation of rights. Libertarians invoke the distinction between justice and charity to reply to those who complain that a Lockean theory of property rights legitimizes crushing poverty for millions. They stress that while justice demands that we not be *forced* to contribute to the well-being of others, charity requires that we help even those who have no *right* to our aid.[6]

III. IMPLICATIONS FOR HEALTH CARE

Now that we have a grasp of the main ideas of three major theories of justice, we can explore briefly some of their implications for health care. To do this we may confront the theories with four questions:

(1) Is there a right to health care? (If so, what is its basis and what is its content?)

(2) How, in order of priority, is health care related to other goods, or how are health care needs related to other needs? (If there is a right to health care, how is it related to other rights?)

(3) How, in order of priority, are various forms of health care related to one another?

(4) What can we conclude about the justice or injustice of the current health care system?

In some cases, as we shall see, the theories will provide opposing answers to the same question; in others, the theories may be unhelpfully silent.

We have already seen that the Utilitarian position on rights in general is complex. If by a right we mean a right in the strict sense, i.e., a claim which takes precedence over mere appeals to utility at all levels, including the most basic institutional level, then Utilitarianism denies the existence of rights in general, including the right to health care. If, on the other hand, we mean by right a claim that takes precedence over mere appeals to utility at the level of particular actions or at some institutional level short of the most basic, but which is justified ultimately by appeal to the utility of the total set of institutions, then Utilitarianism does not exclude, and indeed may even require rights, including a right to health care. Whether or not the total institutional array which maximizes utility will include a right to health care will depend upon a wealth of *empirical facts* not deducible from the principle of utility itself. The nature and complexity of the relevant facts can best be appreciated by considering briefly the bearing of Utilitarianism on questions (2) and (3). A utilitarian system of (derivative) rights will pick out certain goods as those which make an especially large contribution to the maximization of utility. It is reasonable to assume, on the basis of empirical data, that health care, or at least certain forms of health care, is among them. Consider, for example, perinatal care, broadly conceived as including genetic screening and counseling (at least for special risk groups), prenatal nutritional care and medical examinations for expectant mothers, medical care during delivery, and basic pediatric services in the crucial months after birth. If empirical research indicates (1) that a system of institutional arrangements which maximizes utility would include such services and (2) that such services can best be assured if they are accorded the status of a right, with all that this implies, including the use of coercive sanctions where necessary, then according to Utilitarianism there is such a (derivative) right. The strength and content of this right relative to other (derivative) rights will be determined by the utility of health care as compared with other kinds of goods.

It is crucial to note that, for the utilitarian, empirical research must determine not only whether certain health care services are to be provided as a matter of right, but also whether the right in question is to be an *equal* right enjoyed by all persons. No commitment to equality of rights is included in the utilitarian principle itself, nor is there any commitment to equal distribution of any kind. Utilitarianism is egalitarian only in the sense that in calculating what will maximize utility each person's welfare is to be included.

Utilitarian arguments, sometimes based on empirical data, have been advanced to show that providing health care free of charge as a matter of right would encourage wasteful use of scarce and costly resources because the individual would have no incentive to restrain his 'consumption' of health care. The cumulative result, it is said, would be quite disutilitarian: a breakdown of the health care system or a disastrous curtailment of other basic services to cover the spiralling costs of health care. In contrast (proponents of this argument continue) a *market* in health care encourages 'consumers' to use resources wisely because the costs of the services an individual receives are borne by that individual.

On the other side of the utilitarian ledger, empirical evidence may be marshalled to show that the benefits of a right to health care outweigh the costs, including the costs of possible over-use, and that a market in health care would not maximize utility because those who need health care the most may not be able to afford it.

Similarly, even if there is a utilitarian justification for a right to health care, empirical evidence must again be presented to show that it should be an equal right. For it is certainly conceivable that, under certain circumstances at least, utility could be maximized by providing extensive health care only for some groups, perhaps even a minority, rather than for all persons.

Utilitarians who advocate a right to health care often argue that this right, like other basic rights, should be equal, on the basis of the assumption of diminishing marginal utility. The idea, roughly, is that with respect to many goods, including health care, there is a finite upper bound to the satisfaction a person can gain from being provided with additional amounts of the goods in question. Hence, if in general we are all subject to the phenomenon of diminishing marginal utility in the case of health care and if the threshold of diminishing marginal utility is in general sufficiently low, then there are sound utilitarian reasons for distributing health care equally.

Finally, it should be clear that for the utilitarian the issue of priorities within health care, as well as that of priorities between health care and other goods, must again be settled by empirical research. If, as seems likely, utility maximization requires more resources for prevention and health maintenance rather than for curative intervention after pathology has already developed, then this will be reflected in the content of the utilitarian right to health care. If, as many writers have contended, the current emphasis in the U.S. on high technology intervention produces less utility than would a system which stresses prevention and health maintenance (for example through stricter control of pollution and other environmental determinants of disease), then the utilitarian may conclude that the current system is unjust in this respect. Empirical data would also be needed to ascertain whether more social resources should be devoted to high- or low-technology intervention: for example, neonatal intensive care units versus 'well-baby clinics'. These examples are intended merely to illustrate the breadth and complexity of the empirical research needed to apply Utilitarianism to crucial issues in health care.

Libertarian theories such as Nozick's rely much less heavily upon empirical premises for answers to questions (1)–(4). Since the libertarian is interested only in preventing violations of libertarian rights, and since the latter are rights against certain sorts of interferences rather than rights to be provided with anything, the question of what will maximize utility is irrelevant. Further, any effort to implement any right to health care whatsoever is an injustice, according to the libertarian.

There are only two points at which empirical data are relevant for Nozick. First, whether or not any current case of appropriation of hitherto unheld things satisfies the Lockean Proviso is a matter of fact to be ascertained by empirical methods. Second, empirical historical research is needed to determine what sort of redistribution for the sake of rectifying past injustices is necessary. If, for example, physicians' higher incomes are due in part to government policies which violate libertarian rights, then rectificatory redistribution may be required. And indeed libertarians have argued that two basic features of the current health care system do involve gross violations of libertarian rights. First, compulsory taxation to provide equipment, hospital facilities, research funds, and educational subsidies for medical personnel is literally theft. Second, some argue that government enforced occupational licensing laws which prohibit all but the established forms of medical practice violate the right to freedom of contract [3]. Those who raise this second objection also usually argue that the function of such laws is to secure a monopoly for the medical establishment while sharply limiting the supply of doctors so as to keep medical fees artificially high. Whether or not such arguments are sound it is important to note that Libertarianism is not to be confused with Conservatism. A theory which would institute a free market in medical services, abolish government subsidies, and reduce government regulation of medical practice to the prevention of injury and fraud and the enforcement of contracts has radical implications for changing the current system.

Libertarianism offers straightforward answers to questions (2) and (3). Even if it can be shown that health care in general, and certain forms of health care more than others, are especially important for the happiness or even the freedom of most persons, this fact is quite irrelevant from the perspective of a libertarian theory of justice, though it is no doubt significant for the libertarian concerned with charity or other virtues which exceed the requirements of justice. Nozick and other libertarians recognize that a free market in medical services may in fact produce severe inequalities and that there is no assurance that all or even most will be able to afford adequate medical care. Though the humane libertarian will find this condition unfortunate and will aid those in need and encourage others to do likewise voluntarily, he remains adamant that no one has a right to health care and that hence none may be rightly forced to aid another.

According to Rawls, the most basic questions about health care are not to be decided either by considerations of utility nor by market processes. Instead they are to be settled ultimately by appeal to those principles of justice which would be chosen in the original position. As we shall see, however, the implications of Rawls's principles for health care are far from clear.[7]

No principle explicitly specifying a right to health care is included among Rawls's principles of justice. Further, since those principles are intended to regulate the basic structure of society as a whole, they are not themselves intended to guide the decisions individuals make in particular health care situations, nor are they themselves even to be applied directly to health care institutions. We are not to assume that either individual physicians or administrators of particular policies or programs are to attempt to allocate health care so as to maximize the prospects of the worst off. In Rawls's theory, as in Utilitarianism, the rightness or wrongness of particular actions or policies depends ultimately upon the nature of the entire institutional structure within which they exist. Hence, Rawls's theory can provide us with fruitful answers at the micro level only if its implications at the macro level are adequately developed.

If Rawls's theory includes a right to health care, it must be a right which is in some way derivative upon the basic rights laid down by the Principle of Greatest Equal Liberty, the Principle of Equality of Fair Opportunity, and the Difference Principle. And if there is to be such a derivative right to health care, then health care must either be among the primary goods covered by the three principles or it must be importantly connected with some of those goods. Now at least some forms of health care (such as broad services for prevention and health maintenance, including mental health) seem to share the earmarks of Rawlsian primary goods: they facilitate the effective pursuit of ends in general and may also enhance our ability to criticize and revise our conceptions of the good. Nonetheless, Rawls does not explicitly list health care among the social primary goods included under the three principles. However, he does include wealth under the Difference Principle and defines it so broadly that it might be thought to include access to health care services. In "Fairness to Goodness" Rawls defines wealth as virtually any legally exchangeable social asset; this would cover health care 'vouchers' if they could be cashed in or exchanged for other goods ([7], p. 540).

Let us suppose that health care is either itself a primary good covered by the Difference Principle or that health care may be purchased with income or some other form of wealth which is included under the Difference Principle. In the former case, depending upon various empirical conditions, it might turn out that the best way to insure that the basic structure satisfies the Difference Principle is to establish a state-enforced right to health care. But whether maximizing the prospects of the worst off will require such a right and what the content of the right will be will depend upon what weight is to be assigned to health care relative to other primary goods included under the Difference Principle. Similarly, a weighting must also be assigned if we are to determine whether the share of wealth one receives under the Difference Principle would be sufficient both for health care needs and for other ends. Unfortunately, though Rawls acknowledges that a weighted index of primary goods is needed if we are to be able to determine what would maximize the prospects of the worst off, he offers no account of how the weighting is to be achieved.

The problem is especially acute in the case of health care, because some forms of health care are so costly that an unrestrained commitment to them would undercut any serious commitment to providing other important goods. Thus, it appears that until we have some solution to the weighting problem Rawls's theory can shed only a limited light upon the question of priority relations between health care and other goods and among various forms of health care. Rawls's conception of primary goods may explain what distinguishes health care from those things that are not primary goods, but this is clearly not sufficient.

Perhaps because he is aware of the exorbitant demands which certain health care needs may place upon social resources, Rawls stipulates that the parties in the original position are to choose principles of justice on the assumption that their needs fall within the 'normal range' ([9], pp. 9–10). His idea may be that the satisfaction of extremely costly special needs for health care may not be a matter of justice but rather of *charity*. If some reasoned way of drawing the line between 'normal' needs which fall within the ambit of principles of justice and 'special' needs which are the proper object of the virtue of charity could be developed, then this would be a step toward solving the priority problems mentioned above.

It has been suggested that the Principle of Equality of Fair Opportunity, rather than the Difference Principle, might provide the basis for a Rawlsian right to health care ([2], pp. 16–18). While I cannot accord this proposal the consideration it deserves here, I wish to point out that there are four difficulties which make it problematic. First, priority problems still remain. For now we are faced with the task of assigning a weight to health care relative to those other factors (such as education) which are also determinants of opportunity. Further, since the Principle of Equality of Fair Opportunity is lexically prior to the Difference Principle, we must again face the prospect that commitment to the former principle might swallow up social resources needed for providing important goods included under the latter.

Second, because it refers only to opportunities for occupying social *positions* and *offices*, rather than to opportunities in general, the Principle of Equality of Fair Opportunity might be thought too narrow to provide an adequate foundation for a right to health care. Rawls might respond either by defining 'positions' rather broadly or by arguing that opportunities for attaining positions and offices are related to opportunities in general in such a way that equality in the former insures equality in the latter.

Third, and more importantly, Rawls's Principle of Equality of Fair Opportunity takes 'abilities' and 'skills' as given, requiring only that persons with equal or similar abilities and skills are to have equal prospects of attaining social positions and offices. Yet clearly inequalities in health care can produce severe inequalities in abilities and skills. For example, poor nutrition and medical care during gestation can result in mental retardation, and many health problems hinder the development of skills and abilities. Hence it might be argued that if the Principle of Opportunity is to provide an adequate basis for a right to health care it must be reformulated to capture the crucial influence of health care or the lack of it upon individual development.

Each of the theories of justice under consideration offers a theoretical basis for answering some basic questions concerning justice in health care. We have seen, however, that none of them provides unambiguous answers to all of the questions and that each depends for its application upon a wealth of empirical premises, many of which may not now be available. Each theory does at least rule out some answers and each supplies us with a perspective from which to pursue issues which we cannot ignore. Nonetheless, almost all of the work in developing an account of justice in health care remains to be done.[8]

University of Minnesota
Minneapolis, Minnesota

NOTES

[1] In this essay I shall be concerned for the most part with utilitarianism at the institutional level, and I shall proceed on the assumption that a set of institutions which maximizes utility will include rules which bar other direct applications of the principle of utility itself. Consequently, I will mainly be concerned with Rule Utilitarianism, rather than Act Utilitarianism (the latter being the view that the rightness or wrongness of a given act depends solely upon whether it maximizes utility). For an original and interesting attempt to show that Act Utilitarianism is compatible with social norms that bar direct appeals to utility, see [10].

[2] Rawls sometimes refers to the 'Principle of Equality of Fair Opportunity' and sometimes to the 'Principle of Fair Equality of Opportunity'. For convenience I will stay with the former label.

[3] The phrase 'worst off' refers to those who are worst off with respect to prospects of the social primary goods regulated by the Difference Principle.

[4] For a detailed elaboration of this point see [1].

JUSTICE: A PHILOSOPHICAL REVIEW

[5] For a fundamental objection to Nozick's invisible hand explanation, see [11].
[6] P. Singer [12], expanding an argument developed earlier by R. Titmuss, argues that the existence of markets for certain goods may in fact undermine the motivation for charity.
[7] See [2].
[8] I would like to thank Earl Shelp and William Hanson for their very helpful comments on an earlier draft of this paper.

BIBLIOGRAPHY

1. Buchanan, A.: 1975, 'Revisability and Rational Choice', *Canadian Journal of Philosophy* **5**, 395–408.
2. Daniels, N.: 1979, 'Rights to Health Care and Distributive Justice: Programmatic Worries', *Journal of Medicine and Philosophy* **4**, 174–191.
3. Friedman, M.: 1962, *Capitalism and Freedom,* University of Chicago Press, Chicago, pp. 137–160.
4. Kant, I.: 1959, *Foundations of the Metaphyics of Morals* (transl. by L. W. Beck), Bobbs-Merrill Co., New York, Part III.
5. Nozick, R.: 1974, *Anarchy, State and Utopia*, Basic Books, Inc., New York.
6. Rawls, J.: 1971, *A Theory of Justice*, Harvard University Press, Cambridge.
7. Rawls, J.: 1975, Fairness to Goodness', *Philosophical Review* **84**, 536–554.
8. Rawls, J.: 1974, 'Reply to Alexander and Musgrave', *Quarterly Journal of Economics* **88** (November), 633–655.
9. Rawls, J.: 1979, 'Responsibility for Ends', Stanford University, Unpublished Lecture.
10. Sartorius, R.: 1975, *Individual Conduct and Social Norms*, Dickenson Publishing Co., Encino, California.
11. Sartorius, R.: 1979, 'The Limits of Libertarianism', in R. L. Cunningham (ed.), *Liberty and the Rule of Law,* Texas A and M University Press, College Station, Texas, pp. 87–131.
12. Singer, P.: 1978, 'Rights and the Market', in J. Arthur and W. Shaw (eds.), *Justice and Economic Distribution,* Prentice-Hall, Inc., Englewood Cliffs, pp. 207–221.

MARTIN P. GOLDING

JUSTICE AND RIGHTS: A STUDY IN RELATIONSHIP

This paper deals with a few aspects of the relationship between the concepts of justice and rights along historical and, occasionally, analytical dimensions. A close connection between these notions is often assumed. In the fifth chapter of *Utilitarianism* [21], for instance, John Stuart Mill purports to explain the stringency of the demands of justice on the grounds that justice is correlated with that part of morality which is concerned with rights. More recently, Gregory Vlastos has maintained that "an action is just if, and only if, it is prescribed exclusively by regard for the rights of all whom it affects substantially" ([28], p. 53). And still more recently, Ronald Dworkin has asserted that our "intuitions about justice" presuppose that "people have rights . . . " ([12], p. xii). Although there is ample historical precedent for such statements, I want to consider some older sources concerning which it may be difficult to establish what the precise connection between these notions is. In some respects we shall be dealing with a special case of a general problem in the methodology of the history of ideas.

Before I turn to my task I want to say a word about the underlying motivation of this study. There is a great deal of interest in justice and rights in ethics and political philosophy today. While philosophy does not necessarily follow the election returns, as was once alleged of the Supreme Court, it is perhaps not surprising that this interest runs parallel to a resurgence of the rhetoric of justice and, especially, rights in the public arena, domestic and international. Study of the ebb and flow of the historical and analytical relationships between these concepts may contribute to an understanding of the significance of the language of 'rights' and the extent of its indispensability to moral and political discourse. This essay, however, makes no pretense to completeness, and only a few suggestions on these difficult subjects will be offered here.

Further, though this essay does not address issues of justice and health care in particular, it is meant to afford a general framework through which they can be appreciated. It would be a serious mistake to believe that issues of justice and health care do not turn on basic issues of justice and rights. I therefore focus my attention here upon issues of justice and rights in general. The essays in the sections that follow will address these questions in the particular context of health care.

Since the relevant material is so vast, perhaps a secure foothold can be obtained by beginning with a definition of 'justice' that had virtual axiomatic status for countless discussions through the Middle Ages to the early modern period and still frequently serves as a primary literary reference. By so doing, we shall be able to quickly raise a crucial analytical question about the relation between rights and justice, and it will also provide an entrée to the historical and methodological issues I wish to consider in some detail. This definition is the very first sentence of Justinian's *Institutes* (6th century) and is attributed to Ulpian, a Roman jurist of the 3rd century: "Justitia est constans et perpetua voluntas ius suum cuique tribuens" [17]. 'Justice is the constant and perpetual will of giving to each his own right (*ius*)'.

This statement immediately raises a number of questions, of which the most crucial for us is the following. Since, as appears, 'justice' is defined in terms of rights, what does the word 'right' (*ius*) mean? This question has historical as well as analytical significance, and it also will lead us into methodological difficulties.

In order to bring all this into clearer focus it will be useful to compare Ulpian's definition to another statement attributed to him, which occurs a few sentences later in the text. "Iuris praecepta sunt haec: honeste vivere, alterum non laedere, suum cuique tribuere". 'The precepts of right (i.e., right conduct) are these: to live honestly, to harm no one, to give to each his own.' The last clause might be called the precept of justice or the justice element in right action, and in contrast to the original definition it characterizes justice as giving to each 'his own', the term *ius* being absent. Does this show that the notion of rights is not necessary to the concept of justice?, can 'justice' be defined without reference to the notion of rights?, does the notion of 'a right' have any special significance? The last question, of course, is the general motivating problem of this study. (It should be pointed out that the terms 'right' and *ius* are ambiguous. We speak of 'what is right' — as in the precepts of right, what is objectively right — and of 'a man's right'. The latter usage is called 'subjective right' by Continental authors. This expression does not mean that personal rights are matters of subjective opinion.)

A brief sketch of the pre-history of the definition and comparison with parallel definitions of 'justice' will reinforce these questions. Though both statements immediately derive from Stoic teaching,[1] the precursor of Ulpian's definition probably is a passage in Plato's *Republic* (331 D), which cites the dictum of Simonides, the poet: "To give what is owed (*ta opheilomena*) to each is just" [23]. It is a small irony of history that Plato rejected this definition. In a way, however, Plato is able to incorporate it into his own position. His more official definition is, "the having and doing of one's own and what belongs to oneself" ([23], 434 A). One who is just in this sense (i.e., minds his own business) will not have what belongs to others or be deprived of his own, as is pointed out. The rulers, who are assigned the conduct of lawsuits (*dikas dikazein*, justicing the justices) will see to this. (Lawsuits in the ideal state?)

Ulpian's definition also has an obviously Aristotelian overlay, the 'constant and perpetual will' which is the equivalent of Aristotle's notion of a virtue as a trained disposition, a character trait developed through practice. In fact it looks like an Aristotelian expansion of a definition given in the *Rhetoric*, I, ix, 7: "Justice is that virtue on account of which everyone has his own (things) in conformity with the law" [2]. In Book V of the *Nicomachean Ethics* [1], Aristotle distinguishes between universal justice, which is the practice of virtue in general toward someone else, and particular justice, which is concerned with fairness in the distribution of honor, wealth, or other community assets and with correcting unfairness in private transactions (e.g., buying and selling, theft, and contumelious treatment). Ulpian's definition does not make clear whether he has any one of these kinds of justice in view.

Definitions that are parallel to Ulpian's can be found in Cicero. His elaborate discussions of justice do not employ these Aristotelian distinctions, though a term that is introduced may indicate an awareness of them. We shall consider one instance, from *De Inventione*, II, 53, 160, which is often quoted in the Middle Ages. "Justitia est habitus animi, communi utilitate conservata, suam cuique tribuens dignitatem" [8]. 'Justice is the condition of the mind which, with the common utility being preserved, gives to each his own worth (*dignitas*)'. The reference to worth (or, perhaps, desert or merit) recalls Aristotle's position that distributive justice involves distribution according to merit, *axion* (or worth, desert). Cicero's definition has interesting features we cannot look at; the important point to notice now is its difference from Ulpian's two statements: 'his own worth' where the others have 'his own right' and (simply) 'his own.' Cicero's definition seems to derive from the same sources, but these three expressions do not appear to be synonymous.

Our survey, thus far, suggests two alternatives to a definition of 'justice' in terms of rights, though it may still be hard to resist the temptation to read 'rights' into all contexts having the notion of justice. But perhaps the temptation should be resisted at times. Let us illustrate the problem with an example, the colloquy between Abraham and God in *Genesis* 18: 16—33. Obviously, we cannot go into the general subject of justice in Scripture; the example is brought for a limited purpose.

This colloquy occurs after three men have come to announce that Sarah will bear a son. The men then rise up and look out toward Sodom, which is to be destroyed. The text continues:

(17) And the Lord said: Shall I hide from Abraham that which I am doing? ... (19) For I have known him to the end that he may command his children and his household after him, that they may keep the way of the Lord, to do *zedakah* and *mishpat* .... (23) And Abraham drew near and said: Wilt thou indeed sweep away the righteous with the wicked? ... (25) That be far from thee to do after this manner, to slay the righteous with the wicked, so that the righteous should be as the wicked; that be far from thee; shall not the judge of all the earth do *mishpat*?

The first question we face in interpreting this passage is that of translation, for in going from one language to another we often introduce terms which carry connotations not borne by the original. The words *zedakah* and *mishpat* can be respectively translated as 'righteousness' and 'justice' or as 'justice' and 'judgment'. (The Septuagint and Vulgate follow the latter: the last clause in verse 19 becomes *poiein dikaisunen kai krisin* and *facere iustitiam et iudiciam*.) In any case, if 'judgment' is the most accurate translation of *mishpat*, the context suggests that just judgment is the issue here. Abraham's complaint in verse 25, somewhat ironical in the light of verse 19, appears to be based on the most certain intuition we have about justice, namely, the injustice of punishing the innocent. (The Aramaic Targum Onkelos in fact translates verses 23 and 25 in terms of 'the innocent' and 'the guilty'.)

May we, as it were, read the concept of rights into this text? To do so would be to interpret Abraham as protesting against the violation of the rights of the innocent; he would be claiming, so to speak, that the innocent have a right not to be punished. Our question bristles with difficulty, however, because there seems to be no term in the Hebrew Bible which is the literal equivalent of our expression 'a right'. And if so, we would be explicating a text in terms of a concept that finds no explicit acknowledgement in it, which would be misleading in the extreme.[2]

Now it might be argued that this attitude rests upon a mistaken assumption: the assumption that absence of a word in a linguistic-culture indicates a lack of the corresponding concept. (As Professor David Daube has said in another connection, "Too much can be made of the absence from early Latin — or Greek or Hebrew — of a term for 'ownership' " [11], p. 38.) Moreover, the concept of justice, arguably, presupposes the concept of rights, as is shown by an analysis of the 'logic' of certain moral terms. And it might be specifically claimed that the ideas of human worth, of desert, and of being a subject to whom duties are owed (if that is what 'his own' implies) require the notion of respect for rights; rights constitute the 'deep theory' underlying the concept of justice.[3]

It is not my purpose to dispute any of this, except to say that sound *historical* methodology will call for caution. What is legitimate as a philosophical procedure, uncovering the presuppositions of moral or other kinds of discourse, is not necessarily legitimate in the history of ideas. What is it to 'have' a certain concept? Just as we must be careful not to interpret Cicero in terms, say, of Kant's ethical theory, though Cicero probably did influence Kant, so must we be hesitant to use the methods of philosophical analysis to attribute concepts where they do not explicitly appear. The temptation to read 'rights' into our text should be resisted.

But perhaps not always. Here is another Biblical example. Sarah has died and Abraham wishes to acquire a plot to bury her. "Give me," he says to the children of Heth, "*aḥuzath* burying-place with you ... " (*Genesis* 23:44). Literally the Hebrew term means 'the possession of', an interpretation adhered to in the Vetus Latina, *possessionem monumenti*. Jerome's Vulgate (4th century) is more legalistic, *ius selpulchri*. Is this misleading? Perhaps it is appropriate to the context (transfer of real property). Soon we shall see what Francisco Suarez, the late-16th century philosophical jurist, does with it.

We still have our original questions. What does the word *ius* (a right) mean in Ulpian's definition, and does it have some special significance over and above 'his own worth (or desert)' and 'his own'? (I am of course assuming, contrary to some scholars, that *ius* is not a post-classical interpolation.) In Justinian's *Digest* (1. 1, 1) Ulpian in fact cites the 'elegant' definition given by the jurist Celsus: *ius est ars boni et aequi* [17] : "*ius* is the art of the good and the equitable." Fritz Schultz calls this "an empty phrase." "Yet," he says, "it is the only definition of *ius* in our books" ([24], p. 136). It is obvious that Celsus's definition is of little help to us, but this is not the main issue. The main issue is whether 'our books' show any acknowledgement of *ius* in the sense in which a right can belong to an individual. Paulus, a contemporary of Ulpian, does in fact say that *ius* is used in a number of senses, and in the sense in which *ius* is what is always equitable and good, *ius* is *ius naturale*, natural right (*Digest*, 1.1, 11). But this is 'right' in the sense of 'what is right', objective right, not 'a right' of the sort that someone can have.

These remarks are not intended to assert that the Roman jurists did not employ *ius* in a subjective sense, for they did. One example out of many possible is Celsus's statement in *Digest*, 44, 7, 51; "Nihil aliud est actio quam ius quod sibi debeatur, iudicio persequendi" [17]. 'An action is nothing else than a right of suing in court for what is due to us.' And we have already noted an instance from the Vulgate, which shows how popular this sense of *ius* was by the 4th century. So in this respect the Romans can be said to have 'had' the concept of a right even if it is true that, contrary to the practice of modern jurisprudence, they did not make the conception of a right the basis of their arrangement of legal doctrines.[4] But employment of the term in the subjective sense was slow in coming. Cicero uses the word over 100 times in his *De Officiis* [9], and except for two or three occurrences *ius* always refers to 'the right' or 'the law' in more cosmic or institutional significations.

What my remarks are intended to assert is that the jurists did not take notice of *ius* in the subjective sense. When they explicitly recognize different senses of 'right', this is not one of them. The concept of rights simply didn't crystallize as a distinctive idea. And it apparently did not for quite a long time.[5]

I now want to briefly trace highpoints in the historical development of the concept of rights. By seeing what became of this concept we might gain some insight into what it *could* have meant in the early sources, as well as some progress toward understanding the special significance of the language of rights. Plainly, we shall be treading on shaky methodological ground. The Roman pronouncements were tremendously important in the developing tradition, whatever they meant to their authors.

It is not surprising that the first signs of explicit recognition, though even then I think it was not complete, may have been exhibited in discussions of Ulpian's definition of 'justice'. Beginning in the 11th century, the mediaeval glossators on the Roman law wrote extensively on the opening section of the *Institutes*, and their comments probably formed the basis of introductory lectures to law students. A main question is the relationship between justice and *ius*, and Hermann Kantorowicz suggests that these writers were troubled by Ulpian's apparent identification of justice with giving everybody their legal rights.[6] Be that as it may, by the middle of the 12th century the prevailing view is that rights derive from justice *tanquam ex fonte rivuli*, just like rivulets from a spring, as Placentinus puts it.[7] This view was sometimes supported by a rather curious reading of Ulpian's definition. The words *ius suum* is referred to *justitia* rather than to *cuique* (each); so that, as a pupil of Bulgarus says, *tribuit justitia ius suum*, justice renders *its* own *ius* (law, in this context?). The term *ius*, therefore, is given a moral meaning, and is indeed identified with *dignitas* or, more often, *meritum* (worth, desert).

This, of course, is to interpret Ulpian along the lines of Cicero's definition, which was mentioned earlier. Placentinus cites Cicero with approval, and he goes on to explain that giving someone his *dignitas* means giving him a reward *si bene meruerit*, if he shall have merited well, and a punishment if he shall have sinned. Ulpian's definition differs only in its emphasis on justice as a virtue, according to him. On this view, then, we do not have two different definitions of 'justice' here, but merely alternative formulations.

The relationship between *ius* and legal rights is brought out well in a statement by Bracton (d. 1268), who probably is relying on one of the glossators. "As for the words *ius suum*, they mean a man's *meritum*, for because of delict or a pact broken, or the like, one is *privatur quis iure suo*, deprived of his right" ([5], p. 23). According to this statement, it may be said that *meritum* (desert) is the *basis* of legal rights, and Ulpian's formula asserts that justice is giving to each what they are morally entitled to have. Whether this was Ulpian's own meaning is still a question to me. I think that we should not attribute any clear recognition of the notion of rights to the Roman sources, anymore than we should take Ulpian's statement (*Digest*, 50, 17, 32) that 'according to natural law, all men are equal', to mean that all men have equal natural rights, as has often been done.

There is, in any case, one strange aspect to Bracton's discussion, an aspect that probably replicates the treatment of his mediaeval sources. Immediately after the above account, Bracton states that *ius* has a number of different significations. He lists over a dozen — but does not include *ius* as *meritum* among them! I take this as indicating that the concept of rights still had not entirely crystalized as a distinctive notion. Soon we shall look at a Renaissance writer who, though he operates with a rather different conception, is careful to include *ius* in the subjective sense in his list of the meanings of the term.

In any event, it is in the glossators on the Roman law that we apparently find the first glimmerings of an explicit distinction between moral and legal rights. And the conception of a moral right — if we may use this expression despite the fact its literal equivalent does not appear — has two interesting features. First, rights are said to flow from justice. The accompanying explanations make it as clear as our sources ever could that rights do not comprise the 'deep theory' that underlies the concept of justice. The position is, rather, that moral (subjective) rights are grounded on *the* right, in an objective sense. The concept of justice is not 'rights-based,' to use another of Professor Dworkin's terms, but 'right-based'. The individualism which is presupposed by such later accounts had not yet emerged for the glossators. The second feature is that these rights correspond to what modern analytical jurists were to call 'claim-rights' rather than to 'liberty-rights'. The root notion is that of an interest or, more generally, welfare. The basic idea is that of entitlement to a good, rather than freedom, a view which rose to prominence somewhat later. (Of course the glossators did not press their analysis very far; it was left to the 19th century to develop the interest theory of rights. They had all the pieces, only they didn't put them together. In fact they could have found them strewn about in Cicero.)

So much for the lawyers. What about the philosophers? We can briefly consider just a few examples, each significant for a special point. Let us begin with the most important mediaeval discussion of justice, that of Thomas Aquinas (1224–1274). Aquinas, I believe, well illustrates my earlier remark that philosophy does not necessarily follow the election returns.

Aquinas [3] discusses in detail all the distinctions marked by Aristotle and very much more in Questions 57–79 of *Summa Theologiae*, II–II. In Q. 58 he approves of Ulpian's definition, and says that it is pretty much the same as Aristotle's ([1] V, 5, 1134a1): "Justice is a habit according to which a person is said to be active by choosing that which is right," as Aquinas puts it. Aquinas does not concern himself with the problem about *ius suum* which bothered the glossators, and *ius* is not defined as *meritum*. In many places the term *ius* simply is dropped, possibly as a result of the formulation in Ulpian's 'three precepts of right' which merely speaks of giving to each 'his own'. Cicero is cited, but not his definition of 'justice'.[8]

Aquinas had no explicit concept of rights and did not use the terminology of rights, despite the fact that it was widely employed in the 13th century. *Magna Carta* (1215), for example, states that "the English Church shall be free and shall have its rights undiminished, *et habeat iura sua integra* ... " [20]. Aquinas's interpreters have not been deterred, however. Some of them have attributed to him a doctrine of natural rights as well as a concept of rights. So when Aquinas says (Q. 66) that private possession *non est contra ius naturale*, is not opposed to natural right, these interpreters have him affirming a natural right to private property. This tendency did not begin with the 20th century. Leonhardus Lessius (1554–1623), a German commentator on Aquinas, mentions *potestas legitima*, legitimate power, as one of three meanings of *ius*, and interprets Aquinas accordingly. *Meritum*, incidentally, is not listed.[9]

Is violence done to Aquinas's thought if his theory of justice is expounded in terms of the concept of rights? After all, this mode of exposition does seem permissible in the case of legal systems which lack explicit recognition of the notion. The answer to this question depends, I think, on whether rights-language has special significance within moral and political discourse. If it does, sound methodology calls for caution, and we should take Aquinas seriously as showing the possibility of a 'no-rights' theory of justice. If it does not, no harm is done so long as we keep in mind that anything that can be said with rights-terminology can also be said without it. In any case, we should be careful not to attribute to him a doctrine of natural rights, which meant so many things later on. (It would be nice to know who was the first to say that men *have* natural rights.)

But perhaps Lessius was not entirely off the mark. Aquinas does occasionally use the term *potestas* (power) in his work on justice, and in some later writers this did serve as part of the definition of 'a right'. So Aquinas could be said to have an implicit notion of rights, though it does not play an essential role in his treatment. This conception of rights, however, is very different from the one we identified in our discussion of the glossators. One could speculate that it became the dominant conception in the heyday of natural-rights thinking and the political theory of the 17th and 18th centuries.

Probably the most important, and perhaps most influential, treatment of the concept of rights just prior to this period can be found in Chapter II of the *De Legibus* of Francisco Suarez (written in the 1580's and first published in 1612). He discusses various meanings of *ius*. In the strict sense, he says, *ius* signifies the equity which is due to each individual as a matter of justice. Suarez then proceeds to tell us what *ius* in this sense *names*: "a certain moral power [capacity], *facultas quaedem moralis*, which every man has, either over his own property or with respect to that which is due him." (In various places in his book *potestas* is used instead of *facultas*.) In this sense the term occurs in Scripture as well as law. Suarez cites the Vulgate translation of *Genesis* 23:4 and explains that *ius sepulchri*, right of a burying-place, means *facultas sepliendi*, the power of burying ([26], p. 24).

How shall we characterize this notion of a right? Plainly, it is the concept of rights in the subjective sense, the sense of *ius* in which it can be the possession of an individual, and it is a very different notion from that of a right as desert or worth — *meritum* is not even mentioned as one of the senses of the term. Analysis of Suarez's discussion would show that he recognizes the 'claim' aspect of rights; he distinguishes between rights *in* a thing and rights *to* a thing. But the key conception is that of a right as a *liberty* notion. It is possible that the expression 'moral power' directly refers to some active capacity of man, as the Scriptural example suggests.[10] Some later writers in fact define *ius* as *qualitas moralis activa*, an active moral quality.

Suarez was not the originator of this notion of rights. It can be found in earlier 16th century Spanish writers, but he gave it a clearer exposition. Perhaps it derives ultimately from William of Ockham who, according to Michel Villey, was the first person to offer a definition of *ius* in the subjective sense ([27], pp. 158 ff). The crucial passage occurs in Chapter 65 of *Opus Nonaginta Dierum*, one of Ockham's many attacks against the 'Church of Avignon,' written about 1333. "*Ius poli* [divine right?] is nothing but the *potestas* [power, capacity] to conform to right reason without an agreement; *ius fori* [legal right?] is a *potestas* deriving from an agreement, and sometimes conforms to right reason and sometimes does not" ([22], p. 579).[11] While it may be true that Ockham was the first to define *ius* explicitly in terms of power, he certainly was not the first to employ *ius* in this way, for *potestas* occasionally does some jobs of *ius* in Roman sources as well as in the mediaeval jurists. Villey goes so far as to regard Ockham as the inventor of the idea of subjective rights. If so, Ockham was oblivious to his accomplishment. In his later work, *Dialogus de Potestate Papae et Imperatoris*, Part III, Ockham lists a number of meanings of *ius naturale*, and the above definition does not turn up.

In the 17th and 18th centuries expositions of the concept of rights can be found that are more elaborate than that of Suarez's, while in some cases his influence is still evident. The dominant notion of this period is the liberty conception. Hugo Grotius, for instance, in his important work *De Jure Belli ac Pacis* (1625) distinguishes various kinds of rights, all in terms of the idea of *potestas*, and defines the transfer of rights as *alienatio particulae nostrae libertatis*, alienation of part of our liberty (2, 11, 4) [16]. The high-watermark of this conception is probably achieved by William Blackstone [4] in his *Commentaries on the Laws of England* (1765 et seq.), where rights are identified with "that *residuum* of natural liberty, which is not required by the laws of society to be sacrificed to the public convenience" (I, 129).

We have traced the historical development of two very different conceptions of rights. A more refined analysis — which would show their relationships to such notions as positive and negative rights and to ideas employed in modern jurisprudence — is possible, but I shall not undertake such an analysis here. I shall also not attempt to survey theories of justice elaborated by Suarez or other later writers. Whatever the situation in these thinkers, it would be anachronistic in the extreme to read the liberty notion of rights into Ulpian's definition of 'justice'. Ulpian, of course, may be left to take care of his own problems. We are still left with analytical and normative questions as to the relationship between justice and rights.

Which of these is the more fundamental idea? This sorts itself out into subsidiary questions along the lines of positions we considered: a no-rights theory of justice, and theories that employ a merit notion of rights (which, I would argue, actually is one form of a welfare-right conception) or a liberty notion — or even some kind of right we have not considered, e.g., the right to be treated as an equal, which Professor Dworkin favors. Which type of right is the more basic kind? My own inclination would be to support the primacy of the welfare type, but I cannot argue for that here.[12] It is not easy to integrate liberty-rights into a theory of justice, I believe, though they obviously are important. In any case, much depends on the special significance — if any, in the last analysis — that rights-language has in moral and political discourse. This essay has not determined what that special significance is, but perhaps it has cleared the way for a direct confrontation with the subject.

All the classical and early modern writers we discussed do agree on one point: To the extent that they acknowledge rights at all, they agree that rights derive from justice or, to put it in other words, that subjective rights must be subordinated to the objectively right. It could be said that the emergence of a liberty notion of rights throws this assumption into question. Indeed, there are writers who reject the traditional view. The 'epoch-making' significance of Thomas Hobbes's political philosophy, for example, has been ascribed to the fact that Hobbes severed the 'right of nature', which is a liberty notion, from the 'law of nature', and in fact subordinated the latter to the former.[13] The hyper-modern version of this position, endemic to contemporary culture, is that someone can have a right to do what is wrong. The development of the liberty notion of rights allows us to contemplate the possibility of this proposition, as paradoxical as it sounds. One trusts, however, that this does not exhaust the special significance of the terminology of 'rights'.[14]

Duke University
Durham, North Carolina

NOTES

[1] Cicero, who refers to the Stoics, is suggested as the literary source, see ([24], p. 136). All these definitions, according to Schulz, are quite valueless as a characterization of Roman jurisprudence.

[2] The question of 'rights' in the Hebrew Bible is actually quite complicated, only in part because of the multiple meaning of certain terms. On another occasion I hope to

return to this subject and also to the problem of rights-language in Talmudic and mediaeval Rabbinic jurisprudence.

[3] See ([12], Chapter 6: 'Justice and Rights').

[4] See ([6], p. 2).

[5] Cf. ([13], pp. 100 ff, 372 ff). Gewirth argues against the negative view that restricts the concept of rights to the modern era, and he argues that rights are at least implicit in ancient materials. Depending on what he means by 'implicit' I agree on both points. It is my view that the concept does not receive *explicit* recognition by legal theorists and philosophers until the Middle Ages, and that this is a philosophically significant fact. See also [27]. Villey's work I regard as extremely important though I disagree with some of his interpretations and conclusions. My brief account in [15], I would now modify in various respects.

[6] See ([19], p. 62 and, generally, pp. 59 ff., 133 ff., 270 ff).

[7] Cited in ([7], p. 10).

[8] Aquinas does use *dignitas* in his discussion of distributive justice. See ([3], II–II, 61, 2).

[9] See ([18], p. 155).

[10] For a discussion of various interpretations of 'moral power' in Suarez and of the debated question of his relationship to Thomism, see [10].

[11] Villey seems not to have noticed this passage.

[12] For steps in that direction, see [14].

[13] See ([25], pp. 155 f).

[14] I wish to thank my son, Joshua, for his assistance in preparing this essay; he rightfully disclaims responsibility for any of its errors.

BIBLIOGRAPHY

1. Aristotle: 1947, *Nicomachean Ethics* (Loeb Classical Library), Harvard University Press, Cambridge, Mass.
2. Aristotle: 1967, *The 'Art' of Rhetoric* (Loeb Classical Library), Harvard University Press, Cambridge, Mass.
3. Aquinas, T.: 1975, *Summa Theologiae*, Vols. 37–38, McGraw-Hill, New York.
4. Blackstone, W.: 1825, *Commentaries on the Laws of England*, T. Cadell and J. Butterworth, London.
5. Bracton, H.: 1968, *On the Laws and Customs of England*, Vol. 2, G. E. Woodbine and S. E. Thorne (eds.), Harvard University Press, Cambridge, Mass.
6. Buckland, W. W.: 1908, *The Roman Law of Slavery*, Cambridge University Press, Cambridge.
7. Carlyle, R. W., and Carlyle, A. J.: n. d., *A History of Mediaeval Political Theory in the West*, Vol. 2, Barnes and Noble, New York.
8. Cicero: 1960, *De Inventione* (Loeb Classical Library), Harvard University Press, Cambridge, Mass.
9. Cicero: 1968, *De Officiis* (Loeb Classical Library), Harvard University Press, Cambridge, Mass.
10. Composta, D.: 1957, *La 'Moralis Facultas' nella Filosofia Giuridica di F. Suarez* (Pontificium Athenaeum Salesianum, Facultas Philosophiae, No. 43), Societa Editrice Internazionale, Turin.

11. Daube, D.: 1979, 'Fashions and Idiosyncracies in the Exposition of the Roman Law of Property', in A. Parel and T. Flanagan (eds.), *Theories of Property*, Waterloo, Canada, pp. 36–50.
12. Dworkin, R.: 1977, *Taking Rights Seriously*, Harvard University Press, Cambridge, Mass.
13. Gewirth, A.: 1978, *Reason and Morality*, Chicago University Press, Chicago.
14. Golding, M. P.: 1968, 'Towards a Theory of Human Rights', *The Monist* 52, 521–549.
15. Golding, M. P.: 1978, 'The Concept of Rights: A Historical Sketch', in E. and B. Bandman (eds.), *Bioethics and Human Rights*, Little, Brown and Co., Boston, pp. 44–50.
16. Grotius, H.: 1925, *De Jure Belli ac Pacis* (trans. as *The Law of War and Peace*, by F. W. Kelsey), Bobbs-Merrill, Indianapolis, Ind.
17. Justinian: 1973, *Institutiones* and *Digesta*, in P. Krueger (ed.), *Corpus Juris Civilis*, Weidmann, Dublin and Zurich.
18. Kaltenborn, C.: 1848, *Die Vorläufer des Hugo Grotius*, Mayer, Leipzig.
19. Kantorowicz, H.: 1969, *Studies in the Glossators of the Roman Law*, with addenda and corrigenda by P. Weimar, Scientia Verlag Aalen, Darmstadt, Germany.
20. *Magna Carta*: 1965, J. C. Holt (ed.), Cambridge University Press, Cambridge.
21. Mill, J. S.: 1963, *Utilitarianism*, in *The Six Great Humanistic Essays of John Stuart Mill*, Washington Square Press, New York.
22. Ockham, W.: 1963, *Opera Politica*, Vol. 2, Manchester University Press, Manchester.
23. Plato: 1953, *The Republic* (Loeb Classical Library), Harvard University Press, Cambridge, Mass.
24. Schulz, F.: 1963, *History of Roman Legal Science*, Clarendon Press, Oxford.
25. Strauss, L.: 1963, *The Political Philosophy of Hobbes*, University of Chicago Press, Chicago.
26. Suarez, F.: 1971, *De Legibus*, Consejo Superior de Investigaciones, Madrid.
27. Villey, M.: 1969, *Seize Essais de Philosophie du Droit*, Dalloz, Paris.
28. Vlastos, G.: 1962, 'Justice and Equality', in R. B. Brandt (ed.), *Social Justice*, Prentice-Hall, Englewood Cliffs, N. J., pp. 31–72.

FREDERICK S. CARNEY

JUSTICE AND HEALTH CARE: A THEOLOGICAL REVIEW

I

Ethical concepts such as justice may also be theological ones. Religion has played a major role in health care throughout the long history of Western society. It has both motivated its adherents to provide health care institutions and contributed to the shaping of these institutions. In some respects, love or benevolence has been the dominant religious motif that has motivated and shaped the institutions of health care, in other respects it has been religious concepts of justice that have performed this role. It therefore seems appropriate to the purposes of this volume to provide a review of the bearing of theological concepts of justice upon health care.

It is important to recognize that ethical concepts such as justice may also be theological ones. This is the case when they function within an orientation or perspective having two specifiable features. First, there is a belief in God or gods — that is, in one or more objects or states of affairs that transcend ordinary human experience of the sort natural science can describe, that evoke trust and commitment, and that are held to possess an exceptionally great, and usually overriding, importance in human experience. Second, there is a belief that the appropriate human response to God or gods includes living justly toward God (or gods) and toward fellow human beings. In monotheistic religion, for example, it is often held either that God is just, and we should imitate Him, or that God wills (through revelation or nature) that we live justly, and we should obey Him.

This understanding of justice as part of theistic orientation is sufficiently comprehensive to include both monotheistic and polytheistic religions, as well as both major world religions (such as Christianity, Islam, Buddhism, Hinduism, and Judaism) and what are sometimes called 'primitive religions'. It can even be applied to such belief systems as Marxism and secular humanism when they are considered 'theologies', as has recently become fashionable in some circles, by virtue of their exhibiting the phenomena of faith (that is, trust and commitment) in valued objects or states of affairs (that is, 'gods').

Nevertheless, in order to make the materials to be considered in this essay manageable, the focus will be on three monotheistic religions – namely, Judaism, Christianity, and Islam. Henceforth the term 'theological review', as stated in the title, will revolve around them. The reason for selecting these three religions is that they have contributed, individually and jointly, far more than any others to the religio-cultural development of the Western world. It can also be noted that these three religions, which are each rooted in the Abrahamic experience of faith, share a common form as 'religions of revelation'. This is to say, each understands certain events in history to be decisive for the interpretation of the meaning of the rest of history. Their adherents appropriately find these designated events, even after hundreds and sometimes thousands of years, as normatively illuminative of their own experiences. The central importance in which this understanding of faith and revelation is held constitutes the dividing line distinguishing Judaism, Christianity, and Islam from the Hindu, Buddhist, Confucian, Greek, Roman, Mayan, and numerous other religions.

Two common characteristics of Judaism, Christianity, and Islam are especially notable for the purposes of this volume. First, each has placed the problem of justice (both divine and human) at or near the center of religious practice and inquiry. Second, each has a long and distinguished history of extensive involvement in the provisions of health care, including the establishment and support of hospitals and clinics, the recruitment and training of medical personnel, and the education of its constituency in the relation of health care to the proximate and ultimate ends of religious and secular life. This is not to say, however, that the first of these characteristics (concern for justice) is the primary cause of the second (proliferation of health care). Admittedly, they have coexisted within the same religious structure of thought and practice. But so have the concern for justice and the pursuit of agriculture. Yet one would hardly claim that agriculture has been largely caused by the concern for justice. What is much more plausible is that a religion's outlook on justice has served as a regulative notion in that religion's involvement with health care. This, as we shall see, has especially been the case in the prophetic demand that the basic human needs of the poor, the fatherless, and the stranger be provided for and that the fiducial qualities of justice (promise-keeping, truth-telling, property-respecting, and so forth) be honored in all human activities including, of course, health care.

The concepts of justice in these three religions relate to a wide range of normative activity. They are far more extensive than are the various notions of distributive justice (both macro and micro) that occupy the center of attention in North America today in the discussion of justice in health care. First, religious concepts of justice often embrace not only matters of obligation (that is, matters pertaining to what ought to be done) of which the public policy interest in distributive justice today is a part, but also matters of virtue or moral character (that is, matters pertaining to what qualities or dispositions of a person, group, or society are commendable or reprehensible), and matters of value (that is, matters pertaining to what objects or states of affairs are good (or evil), and which goods (or evils) are to be preferred to others). Second, even where matters of obligation are alone the focus of attention, justice is understood to pertain not only to obligations of a distributive sort, but also to obligations of a fiducial nature and obligations pertaining to the rectification of distributive and fiducial wrongs. Third, even where distributive justice is itself central, religious concepts of justice are related not merely to guidance for institutional policy (the preoccupation of most of those interested in distributive justice in health care today), but also to the behavior of individual persons and small groups. Fourth, and perhaps most important, these various parts of justice are understood in theological analysis to have a bearing not only on relations among human persons, communities and institutions, but also on relations between the divine and the human, so much so that divine-human relations of justice are considered to be involved in, and in some instances to make very special demands on, human relations among persons, communities, and institutions.

Some of the issues in this all-too-brief introduction are elaborated at greater length in an earlier essay devoted to the relation of theology to bioethics in general [7]. In this essay I shall henceforth attempt to provide from selected materials of Judaism, Christianity, and Islam a review of (1) the grounds of justice, (2) the content of justice, and (3) the transcendence of justice. This review will necessarily be incomplete, and its selection of materials somewhat arbitrary, because of limits of space. For the same reason, the application of these materials to health care will be more suggestive than systematic.

II

The notion that justice has its grounds in a basically religious concern has a long history that goes back into ancient Mesopotamia (roughly the Tigris-Euphrates valley), from which Abraham (so important not only to Judaism, but also to Christianity and Islam) came prior to his arrival in Canaan. In a royal Sumerian hymn the god Utu is said to put justice in the king's mouth "so that the weak should not be left at the mercy of the powerful, so that the powerful should not do according to their will." And the goddess Nanshe, "who searches the heart of the people," is praised in a hymn as the one "who cares for the widow, who seeks out justice for the poorest" ([14], p. 44). In Babylonian religion, the sun-god Shamash is often represented as holding in his right hand a sceptre and a ring as symbols of justice and righteousness. It is he who, as the representative of heavenly and earthly justice, gave Hammurabi his kingly commission to "let justice shine in the land" ([14], p. 58). In an Assyrian version of the Epic of Creation, Assur, the father of the gods and the bestower of kingship, is also the judge and guardian of justice ([14], p. 67). The abundance of such religious materials on justice has led one authority to say that

the concept of justice is deeply imbedded in the ancient Mesopotamian and Egyptian cultures of the third and second millennia B.C. including the Sumerians, the Akkadians, the Babylonians and Assyrians, the Hittites, Ugarit, and others. In fact, it represents the basis and framework of their understanding of world and reality, of their cosmology. More directly, this concept is essentially related to their understanding of reality as world order as such, and to moral and legal human behavior only inasmuch as such behavior is supposed to be in accord with the order of the world. Thus, justice is the world in order, and the world is ordered in justice ([9], p. 4).

Full monotheistic expression is given by Judaism to these incipient notions of justice that have their source in a cosmic ordering making rigorous demands upon human beings. Abraham is said to be chosen by God to be the founder of "a great and mighty nation," with the understanding that he "charge his children and his household after him to keep the way of the Lord by doing righteousness and justice" (*Genesis* 18: 18–19). For "righteousness and justice are the foundations" of God's cosmic throne (*Psalms* 89: 14; 97: 2). This covenant is understood to be renewed by God's deliverance of Abraham's descendents out of Egypt together with His promulgation through Moses of commandments and statutes setting forth the requirements of His justice (*Exodus* 6: 4–8; 20: 1–17; *Deuteronomy* 5: 1–21). As the Torah proclaims, "I am the Lord your God. You shall therefore keep my statutes and ordinances" (*Leviticus* 18: 4–5). Justice, often spelled out in very precise regulations, thus has its source in early Judaism in God's ordering and caring for His world.

With the advent of Christianity, another aspect in the religious expression of justice achieves prominence. Partly under the influence of the Hellenizing of the eastern Mediterranean world and partly as an outworking of Hebraic religion, a profound concern develops for the status of the individual soul before God and man. The requirements of justice as presented in law and commandments are experienced by some devout persons as both rightly demanded by God and at the same time too rigorous to be fully satisfied, leaving the soul with a distressed feeling of cosmic rejection and personal loss of identity for failure to satisfy them. This intense psychic phenomenon presents the occasion for the discovery by Saint Paul of a cosmic acceptance and new personal identity to be obtained not through full compliance to the justice of the law but through faith in a higher or spiritual justice of God's freely given grace (*Romans* 3: 21–26; 4: 1–5; 6: 15–18). Henceforth, this 'justification', or making right, sets sensitive souls free from the guilt of transgression and the shame of failure, while at the same time keeping the justice of the law and the commandments before them as a continuing and valid guide for their behavior ([5], pp. 110–118). It may be profitable to reflect on the analogous manner in which a kind of 'higher justice' functions in health care today when persons who have obviously abused their bodies through bad health practices and may be said therefore to be under judgment for their transgressions and not really to deserve the careful attention of medical personnel, are nevertheless fully accepted for treatment as patients, and sent forth following treatment with the admonition to keep the requirements of good health practices (the law and the commandments) ever before them as a proper guide to future behavior.

Islam also presents justice as a key component of a cosmic ordering. In this instance, Mohammed is understood to have received and recorded in the Koran the direct revelation from God (Allah in the Arabic language) of His will for all peoples. Muslims are "those who submit" to God's will, and thus become participants in this divine ordering of human affairs. "O you who believe, be maintainers of justice, and thus bearers of witness to God, even though it be against yourselves, your parents, or your relatives, and regardless whether the person is rich or poor" (*Koran*, IV, 135). For "surely God enjoins justice" (*Koran*, XVI, 90) and "God loves the just" (*Koran*, XLIX, 9).

Although each of these three religions (Judaism, Christianity, and Islam) had its origin and its formative early years in the communal acceptance of a belief that God had intervened decisively in human history (primarily through Moses, Christ, and Mohammed) to reveal His just will, these revelatory events were not to remain in any of these three religions as the only source of moral claims such as justice upon their adherents. The gradual adoption into each religion of Greek philosophy, if only partial, was to provide nevertheless a complementary and sometimes competing source of insight into the nature of justice and of the grounds of its claims upon the faithful. The impact of those believers who turned to Greek philosophy was two-fold. First, by seeking out and emphasizing rational principles rather than (or along with) historic revelations, they were able to provide alternative constructions of moral reality. This provided the theological analysis of moral notions such as justice with a richer set of materials upon which to pursue its endeavors. Both revelation and reason could now be employed in highly complex ways, and the combinations of revelation and reason that could be offered were almost limitless.

The second impact of Greek philosophy upon these three religions was to shift a good part of the analysis of moral notions from an emphasis upon obligation theory (commandments, statutes, and the like) to a consideration of virtue theory (traits of character, dispositions, and the like). One comes thereby to speak of justice less as an external law that is binding upon persons and groups, and more as an internal characteristic of persons or groups that is worthy of attentive development. This change was perhaps less radical for Christianity than for either of the other two religions because the world in which Christianity had taken its early shape was one already influenced by Greek philosophy. Furthermore, the Christian doctrine that one is made just by grace through faith, rather than through meeting the obligations of the law as in Judaism and Islam, is in some respects a kind of virtue theory, different of course from Greek virtue theories but with major points of contact with them.

Christianity was also the first of the three religions to turn significantly to Greek philosophy, although the work of the Jew Philo of Alexandria obviously preceded this endeavor. And the particular expression to which Christianity initially turned was, as with Philo, neo-Platonism. Clement of Alexandria, Origen, and Gregory of Nyssa in the East, and Augustine in the West provide, each in his own way, interesting examples of how the historic Judeo-Christian religion of revelation, faith, and commandments can be combined with rational principles of neo-Platonism in the conceptualization of justice and with the Greek aretaic concern to determine what is a just person and to seek the realization of such an ideal. Nevertheless, Judaism and Islam also produced theologians of major competence who combined Abrahamic faith with Socratic reasoning, such as Saadya and Ibn Gabriol did for the Jews and Avicenna and Al-Ghazali did for the Muslims.

While those theologians who followed Plato, or one of the various schools of philosophy deriving from him, tended to emphasize the immediacy of knowledge in their attempt to establish rational grounds for such moral concepts as justice, another group of theologians, who generally came considerably later, found in Aristotelianism an alternative route to rational principles of morality. They built their view of justice on the metaphysics of a natural teleology in which human beings move toward their perfection over time by developing their human functions, and justice is conceived as a constituent and necessary part of that perfecting of basic human functions. Such a view of justice as virtue based on a natural teleology was basic to the moral thought of a large number of Aristotelian theologians in the medieval world, and most notably of the Jew Maimonides, the Muslim Averroes, and the Christian Thomas Aquinas. Although they each accepted a natural teleology and a commitment to a virtue theory of justice, they varied considerably about their conceptions of the meaning and role of justice as obligation. Maimonides, for example, wrote that "the perfection of the moral virtues consists in the individual's moral habits having attained their ultimate excellence. Most of the commandments serve no other end than the attainment of such perfection" ([10], Part III, Chapter 54). Thomas Aquinas, on the other hand, held that virtue and obligation are related in such a manner that virtue is the internal principle and law (that is, obligation) the external principle of human acts ([1], Part I, II, questions 49 and 90).

Aside from the impact of Greek philosophy (both Platonic and Aristotelian) on theologizing about justice, there has been one other philosophical matrix that has had a major influence on the theological enterprise. This is the eighteenth-century critical philosophy of Immanuel Kant and the various schools his thought has spawned during the succeeding two centuries. The acceptance of Kant's critique of the conditions of knowledge undercuts both Platonic and Aristotelian claims to the rational derivation of their principles of justice. Likewise, it denies some religious believers' assertions that God can be an object of knowledge and that the justification of moral claims such as justice can be based on the will of an external lawgiver such as God, even if He were an object of knowledge. Rather, it asserts that God is a postulate of reason made necessary by morality rather than morality being derived from knowledge of the existence and will of God. At the same time, Kant wrote regarding God that "I have found it necessary to deny knowledge in order to make room for faith" ([8], p. 29). The epistemological shaking-up Kant brought to theological learning has had the effect of stimulating numerous 'Kantian' solutions to problems of religion and morality, the most important among them in Jewish circles probably being the works of Moritz Lazarus and Hermann Cohen, and in Christian circles of Friedrich Schleiermacher and Albrecht Ritschl. It is noteworthy that Kantianism, when made the center of philosophical reflection by theologians, has had the effect of reversing the influence of Greek philosophy toward the notion of justice as virtue and instead has restored justice to its primary Biblical place in a framework of law and commands (that is, of obligation).

This rehearsal of the sources of justice in the three religions under consideration has made evident the rich, albeit confusing, tradition one needs to take into account if he is to avoid theological parochialism or dogmatism. Most informed persons would agree, I think, that the parameters of responsibly undertaking to represent any of these three religions in their interpretation of justice would include an accountability both to the claim of faith that God has revealed Himself in history as the One who orders the world in justice and the claim of reason to provide a plausible account of the nature and content of that revealing activity and of the appropriate human response to it.

III

The content of justice from the viewpoint of these three religions includes a number of different matters. First, we might look initially at what forms of justice there are. In the early scriptures of Judaism, Christianity, and Islam there are numerous references to matters of trust among persons. Such matters pertain to giving full weight and making correct change in commercial dealings, keeping one's commitments, not putting at risk the neighbor or members of his family or his property, and so forth. Throughout the literature, violations of these fiducial relationships are called acts of injustice, and faithfulness to these relations is acknowledged to be the mark of a just man. Furthermore, in the writings of all three religions there are repeated assertions that the disadvantaged make a claim upon the community and the members thereof. The classes of the disadvantaged that are most often singled out are the poor, those who are without a father or husband, and strangers in the community. While the scriptures seem to be hardly interested at all in general theories of distributive justice, they bear down very heavily upon the injustice of leaving the poor, the fatherless and widow, and the stranger in distress. I take it that this matter can be considered as evidence that the Bible and the Koran have or imply a special theory of distributive justice. Finally, much attention is devoted to the justice involved in correcting both fiducial and distributive wrongs, including both appropriate punishments and compensation. It is interesting that, if the case be granted that the regulatory moral notion of concern for the needs of the disadvantaged is a kind of distributive justice, these scriptures then contain all three of Aristotle's forms of justice as found in his *Ethics*: justice as reciprocity, justice as fairness in distribution, and justice as correction of wrongs ([2], pp. 117–128). That the needs of the disadvantaged, especially the poor, make a claim of distributive justice upon the believer, rather than providing merely an occasion for desireable but not required charity or beneficence, is fairly clear in Islam and Judaism, though somewhat ambiguous in Christianity. In Islam, the 'poor due' or '*zabat*' (the annual distribution of two-and-one-half percent of one's wealth to the poor) is one of the 'Five Pillars of the Faith' which no faithful Muslim with sufficient means is morally or religiously permitted to overlook. In Judaism, '*tsedeq*' (or justice) requires distributions to the poor and needy, and to the orphan and widow. In Christianity, there occur passages in which alms are a work of charity and other passages in which they are a rightful demand of justice, as for example in a morally-influencial work of Gregory the Great: "We discharge our payment of alms more as a duty of justice than we perform it as a work of mercy" ([6], p. 139 – my translation).

Moreover, the relationship between hospitality, a sub-duty of distributive justice especially nourished in the desert and near-desert circumstances of early Judaism and Islam, and the later development in all three religions of long traditions of hospital creation and support deserves notice. When a community of faith acknowledges that provision for the needs of the stranger and the poor is a matter of simple justice, it is only a short step from hospitality to hospitals and from alms to eleemosynary institutions.

A second matter that needs attention is whether in these three religions justice is a sufficient principle for moral guidance in practical activity such as health care. For the most part, the answer is no. It is generally held in these religions that justice (even when inclusive of all its forms: fiducial, distributive, and corrective) needs to be supplemented by mercy or love. However, there are variations among these religions regarding the relation of mercy or love to justice. Judaism tends to maintain a delicate balance between them, both in God's cosmic ordering and in man's duties to his neighbor. There is a Jewish midrash (exegesis) that God has two thrones, one of justice and the other of mercy. If He did not rule from the throne of justice, there would be chaos in the world. But if He did not also ascend the throne of mercy, the world could not endure. Furthermore, the two divine names of Elohim and Yahweh, the first often rendered 'God' and the second 'Lord', were understood to express these two attributes of justice and mercy respectively, and to be jointly necessary to manifest the full character of God. These same attributes become normative for man's relation with his fellow-man by being embodied in the often precise requirements of the covenant relation between God and man.

In Islam, however, the balance seems to be tilted more toward justice than mercy, or at least to be tilted in such a manner that the doing of justice is often a prerequisite to the receiving of mercy. God's "justice is unrelaxing. He will forgive none but those who believe in Him and obey His commandments" ([13], p. 179). As the Koran proclaims, "Observe your duty to Allah that you may obtain mercy" (XLIX, 10; also III, 126; VI, 156; VII, 61; XXIV, 55; XXVII, 47).

Christianity, on the other hand, often seems to tilt the balance the other way. Love is understood to express more fully the nature of God's relation to His world than does justice, even though God is conceived of as being just and as expecting justice of those who are related to Him in love. Nevertheless, the relations between these two concepts in Christianity (as in Judaism and Islam) is much more intricate than can be explicated within the space limitations of this essay, and vary somewhat with different authors. Consider, for example, the following passage by Saint Basil of Caesarea that, in at least one type of circumstance, advocates a lexical priority of justice before mercy.

If you will make an offering to God from the fruits of injustice and rapine, it would be better not to possess such wealth and not to make an offering. . . . You must, therefore, combine justice with mercy, spending in mercy what you possess with justice. . . . If you give alms to the poor after you have despoiled them of their goods, it were better for you neither to have taken nor given. . . . Do not commit injustice on the pretext of offering your mercy to God — mercy made possible by injustice ([4], pp. 508–509).

The third matter in this exposition of the content of justice is the employment of justice as a value concept, and not merely a concept of obligation or of virtue as has thus far been discussed in this essay. Value pertains to the assessment of what things, persons, or states of affairs are important, in what sense they are so, and which are to be preferred to others. A just valuation is one that is made with a conscientious sense of fairness and that does not misrepresent the agathic status of what is under consideration. It is the rendering to each his due in the realm of evaluative perception. This is what the theologian Paul Tillich would seem to have had in mind when he wrote that intrinsic justice is the first level of justice, and described it as follows:

The basis of justice is the intrinsic claim for justice of everything that has being. The intrinsic claim of a tree is different from the intrinsic claim of a person. . . . Justice is first of all a claim raised silently or vocally by a being on the basis of its power of being ([15], p. 63).

The acceptance and employment of such an understanding of justice as value has been at the source of a number of the most important moral developments in the history of Judaism, Christianity, and Islam. It was present in the pronouncements of the Jewish prophet Second Isaiah that God, as creator of the whole world, extended His loving-kindness and salvation to all peoples, thus establishing a universal claim of value for all persons. It was a potent factor in the rescue by Christians of unwanted newborns exposed by their Roman parents on suburban hillsides, and the cherishing and raising of them as valued children of God. It was the basic reason for special status accorded Jews and Christians as 'People of the Book' by Mohammed and his followers. It led a number of Spanish Christian theologians in the sixteenth century to engage in strong advocacy for the intrinsic value of the American Indians, who had often been slaughtered off as possessing only limited instrumental value by conquistadors or enslaved for the same reason by commercial enterprisers. It was the insight upon which American Blacks came to be accepted fully as persons through the civil rights struggles of the nineteen sixties. And it is a prerequisite understanding to the idea of equal access to basic medical care by all persons regardless of race, religion, sex, or economic status, and to the equally important notion that each patient and medical worker has an essential worth that ought to be acknowledged in the operation of health care institutions.

Another aspect of the employment of justice as a value concept pertains to the intrinsic justice involved in the acknowledgement of persons not only as isolated individuals, but also as participants in families, groups, and communities, indeed, as obtaining part of their own sense of identity and purpose through such participation. So it was in the Old Testament understanding of covenant, which was always between God and community (or the representative of a community), never between God and a particular individual, and which provided the most basic sense of identity available to each of the participants therein. So also is it understood by both Christianity and Islam that the appropriate response to God creates a community of the faithful, indeed even with those beyond the faithful, from which community one's most important identity is derived. This value perception of the communal roots of each person is fraught with many implications for the treatment of patients and medical personnel in health-care relations and institutions.

IV

The final topic of this theological review, the transcendence of justice, has three parts. The first is perhaps best expressed by Reinhold Niebuhr. It is the sense in which the principles of justice are both operational norms to be applied to particular situations and at the same time ideals that can never be fully realized in history ([12], vol. II, pp. 247–256). If the principles of justice are equality, liberty, and community (or brotherhood), as has often been suggested, then each of these principles has a normative bearing upon human structures and practices that cannot rightfully be ignored. Loyalty to them is regulative of human behavior and institutions. At the same time, they transcend even the best of intentions and efforts to embody them, and stand in judgment over any claims that their demands have been fully, or even adequately, met. Such claims lead to self-deception and to one or another form of 'ideological taint'.

The second part of the transcendence of justice to be discussed here was made famous by Saint Augustine. He began by recalling that Cicero had defined the state as an assemblage of persons united by an agreement about justice and a sharing of common interests. From this it follows that there cannot be a state without justice. For "remove justice, and what are states but gangs of criminals on a large scale? What are criminal gangs but petty states" ([3], p. 139). Nevertheless, Augustine claimed, if "justice is to render to each his due," as had been asserted by Plato, Cicero, Ulpian, and countless others, and accepted universally, there has never been a state, at least in Cicero's sense. For no state is capable of rendering to God His due. What is due God is full obedience of creatures to their Creator ([3], p. 882). Augustine proposed that a state be defined as an assemblage of persons united by an agreement about the object of their love. In this way he could provide without contradiction for the obvious existence of many states throughout history.

What is significant for the purposes of this essay about Augustine's analysis is his claim that the justice of human acts is not to be found merely in the way they express relations among human beings and institutions, but also in the manner in which they give expression at the same time to important relations between human beings and God. Human acts of justice or injustice thus have a transcendent theological dimension, and this transcendent dimension is in part determinative as to whether they are rightly to be considered just or unjust. Something of this same Augustinian perspective was advanced in the twentieth century by H. Richard Niebuhr, who held that all human responses are properly to be understood as responses to God, and furthermore that "God is acting in all actions upon you. So respond to all actions as to respond to His action" ([11], p. 126). The import of such theological outlooks is to interpret human acts, and by implication human dispositions and attitudes as well, as having a divine-human point of reference and not to be sufficiently understood simply by a secular analysis.

The third and last part of this discussion of the transcendence of justice pertains to theodicy, or the question of the justice of God, given the seemingly unjust suffering that occurs in human lives. If God is both powerful and good, how is such evil in His world to be explained? There is probably no problem of justice in health care that is more profoundly experienced nor less adequately addressed than this one, perhaps because medical personnel and their allies in philosophy, law, and the social sciences find it enormously complex, frustrating, and even fruitless to consider. But while ignored, it does not go away. A number of solutions have been suggested for it. I shall mention three of the major ones. First, there is the proposal that undeserved suffering be considered 'the afflictions of God's love', a kind of testing and refining of the virtue of the righteous. So thought the Jewish theologian Saadya and the Christian Saint Augustine. Second, such suffering was held to be a punishment for sin, and thus an occasion for self-examination, as both Maimonides and John Calvin proposed. Finally, Immanuel Kant and others held that the lack of correlation in this life between moral behavior and human desert suggested the explanation of life beyond this life in which the moral demand for justice will be transcendently met.

V

Judaism, Christianity, and Islam each have exhibited a concern for justice which has implications for the contemporary discussions about health care. The early theological formulations of distributive justice focused mainly on the care of disadvantaged groups within society whereas later constructions became more developed statements of distributive justice similar to, yet distinct from, philosophical notions. The duty of the creature to the creator to be just in every aspect of human activity, not only in the formal distribution of social benefits and burdens, in part distinguishes the theological from the philosophical tradition.

A transcendent ground, however, does not disqualify the relevance of theological reflections for issues of justice in bioethics. Representatives of these religions can join with others who do not share these beliefs but who share an interest in a more just society, to work together towards a common end, without necessarily forfeiting their distinctive orientation or justification of their efforts. A value of a religious perspective on matters of justice and health care is that it is found by some to provide a more adequate interpretation of human life and events. The proceeds of theological reflections on justice represent for some a more clear and satisfactory vision of truth than those represented by a non-theistic orientation. Theological perspectives are only some among many in a pluralistic society. Yet their influence in public discussion has been significant in the past and probably will continue to be in the future. The distinctive resources of religion can inform and enrich the modern pursuit of the meaning and requirement of justice for health care.

Perkins School of Theology
Southern Methodist University
Dallas, Texas

BIBLIOGRAPHY

1. Aquinas, T.: 1947, *Summa Theologica*, 3 vols. Benziger Brothers, New York.
2. Aristotle: 1962, *Nichomachean Ethics*, Bobbs-Merrill Library of Liberal Arts, Indianapolis.
3. Augustine: 1972, *The City of God*, Penguin Books, Baltimore.
4. Basil of Caesarea: 1950, 'On Mercy and Justice', *Ascetical Works*, Fathers of the Church, Inc., New York.
5. Brunner, E.: 1945, *Justice and the Social Order*, Harper and Brothers, New York.
6. Carlyle, R. W. and Carlyle, A. J.: 1950, *A History of Medieval Political Theory in the West*, Vol. 1, William Blackwood and Sons, Edinburgh.
7. Carney, F.: 1978, 'Theological Ethics', in *Encyclopedia of Bioethics*, Macmillan Free Press, New York, Vol. 1, pp. 429–437.
8. Kant, I.: 1929, *A Translation of Kant's Critique of Pure Reason*, trans. N. Smith, Humanities Press, New York.
9. Knierim, R.: 1979, 'The Biblical Concept of Justice', an unpublished address by Knierim, Professor of Old Testament, Claremont Graduate School, Claremont, California.
10. Maimonides, M.: 1963, *The Guide of the Perplexed*, University of Chicago Press, Chicago.
11. Niebuhr, H. R.: 1963, *The Responsible Self: An Essay in Christian Moral Philosophy*, Harper and Row, New York.
12. Niebuhr, R.: 1949, *The Nature and Destiny of Man*, 2 vols., Charles Scribner's Sons, New York.
13. Rahbar, D.: 1960, *God of Justice: A Study of the Ethical Doctrine of the Qur'an*, E. J. Brill, Leiden.
14. Ringgren, H.: 1973, *Religions of the Ancient Near East*, The Westminster Press, Philadelphia.
15. Tillich, P.: 1954, *Love, Power, and Justice*, Oxford University Press, New York.

LAURENCE B. McCULLOUGH

JUSTICE AND HEALTH CARE:
HISTORICAL PERSPECTIVES AND PRECEDENTS

I. INTRODUCTION

The history of medical ethics is a very old and distinguished one, with its roots in ancient Greece and its latest expressions appearing regularly in learned journals and the meetings of professional societies. Because it is so old, spanning more than a score of centuries, the history of medical ethics cannot be easily summarized nor can it be captured in altogether reliable generalizations. Nevertheless, there do seem to be persistent leitmotifs to that history. Two in particular stand out. The first is an emphasis on what Robert Veatch has called Hippocratic medical ethics, a focus on the ethical dimensions of the patient-physician relationship [34]. Thus, matters concerning confidentiality and the communication of information to dying patients, for example, are addressed frequently. The second common theme is that of collegial relationships. Here the concern is for the proper bounds of relationship among physicians, or between physicians and other health care providers, for example, surgeons, apothecaries, irregular or unorthodox practitioners, and the like. We find this twin set of concerns in writings as diverse in time, presupposition, and cultural context as the Hippocratic corpus [19] and Thomas Percival's *Medical Ethics* [27]. What is missing from this history, certainly before the eighteenth century, is a concern for ethical issues in medicine of a broader, more institutional character.

It is interesting, I think, to speculate on why this is so. A plausible explanation may be the following. It has really not been until the last two centuries that medicine acquired a solid and reliable scientific foundation for the study and treatment of disease. One of the consequences of the rise of scientific medicine has been the increasing ability of medicine to affect matters of health that are outside the patient-physician relationship. That is, not only did physicians become increasingly adept at the detection and treatment of diseases of individual patients, they also became skilled in understanding and managing disease conditions and patterns in populations. Thus, as we shall shortly see, explicit and systematic concern for matters such as public health finds its way into the history of medical ethics coincident with the early formation of modern medicine in the eighteenth century.

In more recent times, these concerns have been buttressed in another manner. As medicine — in our century especially — has assimilated an increasingly powerful science and its attendent technology, its ability to affect our lives, for better or worse, has increased. Hence, expectations of the public concerning the broader application of medical knowledge and skills have increased. In addition, the willingness within the medical professions to use medical science for the public good has also increased. Indeed, at times in the history of medical ethics during and before this century it has been thought to be obligatory — from perspectives both within and without medicine — for medicine to serve the public good. In summary, then, one would say that our time is not the first in which there has been a serious concern for and inquiry into what might be called institutional medical ethics.

Now, recent years have witnessed recrudescence of such inquiry. While contemporary efforts in this area have not been many, the discussion has nevertheless been quite vigorous. Before addressing the historical dimensions of inquiry into justice and health care, it will be useful to provide a topography of the present scene. The point in doing so is to develop categories of thought, which can then be placed in historical perspective. Thus, the historical inquiry undertaken here is not meant simply to be of intrinsic interest but aims to uncover historical precedents for contemporary concerns, precedents which may provide a basis for critical evaluation of contemporary efforts, about which some suggestions will be offered.

This essay, then, falls into three parts: The first comprises the promised topography of the contemporary scene. In the second, I examine some historical writings on institutional medical ethics. Here the focus will be on the eighteenth and nineteenth centuries, during which major developments occurred. Finally, I close with suggestions about the manner in which historical inquiries of this kind can contribute to a critical evaluation of inquiries into justice and health care.

II. A TOPOGRAPHY OF THE CONTEMPORARY SCENE

Four major trends seem to have emerged in the recent literature on institutional medical ethics. Not all have received equal development, however, and so the task of fully articulating each remains to be completed. Indeed, some views are more at the stage of suggestions or intimations than they are full-blown accounts. Nevertheless, the following topography is, for present purposes, sufficiently accurate.

All of the views considered here share a common theme: the present health care system is riddled with inequities. Unfair advantages are accruing to individuals or groups of individuals on the basis of such factors as race, economic status, geographical location, and age. In short, the present arrangements for the organization and delivery of and access to health care in our society are unjust and in need of reform. The various accounts of the proper foundations for reform fall into the following classifications:

Rights to Health Care

A traditional response in Western thought to injustice in social institutions is the appeal to basic rights. That is, the injustice in question should be understood primarily in terms of a violation of certain, specifiable, and basic entitlements of persons. In the present context, the injustice involves blocking access simply on the basis, say, of economic status, race, or geographic origin. Access to the benefits of social institutions should not be based on such considerations, it is argued, but should be granted as an entitlement, as a right lodged against those selfsame institutions. The right to health care, it would seem, obtains as a *basic* right: it originates independently of considerations of utility, property, professional prerogatives, or the willingness of society to underwrite the costs of services required to satisfy this right [26].

The emphasis in discussions of the right to health care is two-fold. Some challenge directly whether such a right exists at all, arguing that it does not because it is in contradiction to the basic rights of medical practitioners [31]. Others grant that there is a right to health care and focus on the scope of that right. A consensus seems to have emerged that the right is quite limited, securing access only to a decent, fair, minimum level of health care ([8, 12]). A sub-set of this school is one which holds that health care as a right must be understood in institutional contexts. That is, abstract claims to basic rights will not do. Whatever rights one has, one comes to possess in virtue of being within an institutional setting and only in this way [5]. Thus, one possesses the right to health care insofar as institutions confer such a right.

Utilitarian Theories of Health Care

The next three views share a common presupposition. H. Tristram Engelhardt captures this presupposition nicely when he points out that, for many of the views on justice and health care, health care as a right emerges from "concerns for particular goods and values." [10] That is, one must first get clear on what a just social order for health care would be like and then create, not discover, rights accordingly.

A utilitarian approach is, in this respect, rather straightforward. Reasonable people should be able to agree on the goods, i.e., health and health care, to be pursued and on the means to maximize those goods for the greatest number of people in need of them. Some approach the issues in terms of a pure utilitarian account [11], while others — in a more reasoned and persuasive manner — argue along Rawlsian lines [15].

Interestingly, developments in the People's Republic of China can be understood along these same lines ([3, 32]). The Chinese, if we understand them correctly, can be taken as having rather explicitly placed an emphasis on the value of preventive medicine and health maintenance. In this way, presumably, the common good of all and of the state is served. A sturdy and large population has been developed to provide the human resources necessary to the making of a nation, a goal that would not have been achieved if, let us say, resources had been devoted to high-technology, tertiary-care medicine, whose benefits devolve upon a far smaller proportion of a population.

Justice and Health Care

An approach related to but still distinct from the utilitarian view centers on justice and health care. Retributive and distributive theories of justice are applied to health care access, provision, and distribution. Proponents of this sort of approach tend to be skeptical about rights theories in general and about the right to health care in particular. Rather, rights emerge, if at all, only in virtue of an account of justice.

Two distinctive views may be noted. One, defended by Tom Beauchamp and Ruth Faden [4], holds that a right to health care emerges from considerations of scarce health resources, limited by fairly strict cost/benefit considerations. A focus on the right, *simpliciter*, to health care, fails to appreciate that there is a more fundamental concern:

... if there is a right to health care goods and assistance it is only because there already exists an obligation to allocate resources for goods and assistance. We conclude that the major issues about rights to health and to health care turn on the justification of social expenditures rather than on some notion of natural, inalienable, or preexisting rights ([4], p. 130).

The second view is advanced by Robert Veatch [37], who integrates considerations of justice and rights. The former is the primary context for his argument, however. As he puts it, "The primary assumption is that a society must evaluate its institutions using a basic set of principles," ([35], p. 172). Those principles, it would seem, are just-making considerations about resource allocation. A right to health care derives from those principles of just allocation of health resources. Interestingly, Veatch's formulation of the right is more encompassing than that of Beauchamp/Faden or adherents of the first school of thought.

Justice requires everyone has a claim to health care needed to provide an opportunity for a level of health equal, as far as possible, to other persons' health ([37], p. 134).

Both views, it should be emphasized, share a common strategy. The right to health care does not stand independently from but emerges only within an account of just health care.

Virtue and Obligations to Health Care
A final account of just health care appeals neither to a notion of rights nor to explicit theories of justice. Instead, the appeal is to "what constitutes a decent and humane society" ([9], p. 145). This is James Childress's formulation [9], who adopts Max Weber's language of 'goal-rational' and 'value-rational' conduct. The latter is key here. It "involves matters of virtue, character, and identity that are not easily reduced to ends, effects, or even rules of right conduct" ([9], p. 143). That is, moral conduct is to be motivated and judged by how well our actions express or realize certain values, attitudes, and characters.

Mark Siegler [33], correctly I think, argues that this sort of view is closest to the "traditional covenantal relationship between doctor and patient" ([33], p. 154). Indeed, Siegler's view is that a rights approach and the approaches to inequity in health care through considerations of utility or justice will radically disrupt the proper ethical bounds of the patient-physician relationship, by grounding that relationship in 'contractual minimalism' and/or 'contractual maximalism'. Siegler's view, I take it, is that we should not tamper with what he sees as the traditional covenantal relationship between physicians and their patients (see also [23]). Instead, obligations regarding the organization and delivery of health care are to be made out in terms of a reasonable concept of the virtuous physician. The right to health care, on this view, has no independent standing but is "better justified from the viewpoint of beneficence or virtue . . ." ([33], p. 155).

Interestingly, three of those approaches are not original. Wittingly or not, they derive from and can be placed in the tradition of views that predate them by almost two centuries. An exception may be the view that just-making considerations are the key to understanding institutional medical ethics, though an attentuated version of this view can be found in the eighteenth century. Let us now turn to an examination of historical sources, from the eighteenth and nineteenth centuries, on the ethical dimensions of medicine as a social enterprise. In the course of this examination we shall study precedents for inquiry into the right to health care, utilitarian accounts in the form of theories of medical police, and accounts based on concepts of the virtuous physician.

III. HISTORICAL PERSPECTIVES

I have stated that the focus of historical inquiry here will be on the eighteenth and nineteenth centuries. In selecting this focus I do not mean to claim that an interest in institutional medical ethics first originated in the eighteenth century. While this may well be the case, reliable conclusions cannot be drawn until sustained scholarly investigations into the history of medical ethics have been undertaken. Nevertheless, the eighteenth and nineteenth centuries produced a significant literature on institutional medical ethics. In France there occurred, during the Revolution, debates about the right to health care. In the German states, the concept of medical police was developed to address, in a utilitarian fashion, the broader social dimensions of medicine. Finally, in Great Britain and in North America institutional medical ethics was based on a concept of the virtuous physician. We now turn to an examination of these three schools of thought with a view toward uncovering their philosophical underpinnings and implications.

The Right to Health Care

To the best of my knowledge, the French were the first to articulate what amounts to the right to health and to health care, during the period of the French Revolution. This position contrasts rather sharply with that which was held before the Revolution. Though Meicier in his Utopian work, *L'An deux mille quatre cent quarante* (*The Year 2440*, published in Amsterdam anonymously in 1770), presented the idea that "All citizens have a right to free treatment, and are not driven to the hospital by extreme indigence" ([30], p. 209), the prevailing view was that poverty and the health problems of the poor were a matter of charity. At first addressed on a local level, these problems were conceived eventually as national in nature. Influenced perhaps by the medical police movement, writers like Chamousset (1717–1773) are explicit about the state's responsibilities.

Men are the most valuable possession of a state, and their health is their most valuable possession. But it is not enough that they have the means of perserving it. An object of more importance to them is that in case of sickness they may count on all the aid necessary to their recovery . . . .

JUSTICE AND HEALTH CARE 57

There are asylums available to the destitute, and that is a resource useful to those to whom it is not humiliating to accept the free assistance which charity offers ([30], p. 211).

This view, that provision of health care should be performed as an act of charity, was built on the Catholic tradition of charity, a movement that had already resulted in the establishment of the famous Hôtel-Dieu in Paris. In the period leading up to the revolution this view was challenged. Jacques Turgot (1727–1781), for example, while governor of Limousin in 1770 put the matter this way: "The relief of men who suffer is the *duty* of all, and all the authorities will cooperate toward this end" ([30], p. 213, emphasis added). That is, the provision of health care is not supererogatory but a matter of strict duty.

During the revolutionary period itself, this view was taken still further by the Committees on Health and on Poverty of the States-General. The Duc de la Rochefoucauld-Liancourt, chair of the Committee on Poverty, argued the case in terms of a right to existence, a "fundamental truth of all society" and one of the "rights of man" ([30], p. 231). This fundamental, indeed natural, right has two prongs. The first is a right to be free of poverty, which destroys one's means of support. Thus, public assistance is now an obligation of society founded in the natural rights of its citizens. The second is the right of the sick and needy to health care, articulated by the Committee on Health. Each citizen has a right to "prompt, free, guaranteed and total care" ([38], p. 971), based on the right to existence.

It seems clear that the French revolutionists were influenced by Locke's theory of natural rights [22]. Natural rights in his view may be conceived of in terms of basic entitlements to the necessary conditions of human existence. Locke himself included health among the natural rights. If one could not provide for one's own health, the French position says, one is entitled to health care as a matter of obligation based in each person's right to existence. Thus, neither age, infirmity, illness, nor poverty shall be obstacles to adequate health care ([38], p. 971). Citizens possess natural rights logically, morally, and historically prior to the existence of particular social orders. These rights are the basis upon which societies will be judged just or unjust. Thus, a society that fails to meet obligations of health care to its citizens is clearly unjust, according to the French revolutionists.

Medical Police

A rather different account of obligations regarding health and health care can be found in the writings on medical police. (We are indebted to the scholarly work of George Rosen for our knowledge on this subject [30].) This concept was developed by German thinkers in the eighteenth century. Although in the seventeenth century a number of suggestions were advanced on the state's role in health by, among others, Leibniz ([30], p. 126), the concept of medical police underwent systematic development in the eighteenth century, culminating in the work of Johann Peter Frank.

It was a physician, however, Wolfgang Thomas Rau (1721–1772), who first coined the term medical police, according to Rosen. The background assumptions of the concept are worth noting. The chief motivation for attention to matters of health was the need of the monarch for a large, sturdy, and healthy population. On the one hand, this need was felt in peacetime, to maintain the basis of a strong and growing economy. On the other hand, times of war called for large reserves of persons able to bear arms and the burdens of combat. Rosen summarizes Rau's views on this topic as follows.

... every monarch needs healthy subjects who will be able to fulfill their obligations in peace and war. For this reason, the state must care for the health of the people. The medical profession is obligated not only to treat the sick, but also to supervise the health of the population ([30], p. 138).

The problem at this time, though, was that quacks and charlatans abounded, undoing the good work of physicians for the monarch's and citizens' common cause. Rosen reports that Rau addressed this problem as well.

In order to have competent medical personnel, it is necessary to enact a medical police ordinance which will regulate medical education, supervise apothecary shops and hospitals, prevent epidemics, combat quackery, and make possible the enlightenment of the public ([30], p. 138).

Rau's scheme gives broad scope to the concept of medical police and charges the medical profession with a variety of tasks heretofore not conceived of as obligations.

Frank (1745–1821) provides a systematic account of these distinctive obligations of physicians to the state in his monumental eight-volume work *System einer vollständigen medicinischen Polizey*, published between 1786–1817 [13]. Selected translations of Frank's work have recently been published under the accurate title, *A System of Complete Medical Police*, by Erna Lesky [14]. The latter is an indispensible source for students of the development of the concept of medical police.

Frank takes, as his, the task "to acquaint persons in authority in human society with the necessities of the nature of their subjects and with the causes of their physical ills" ([14], p. 5). Thus, like the French, Frank does not address himself exclusively to medical practitioners, though no doubt he meant to include them. His principal audience was the state. For Frank, of course, the state was the monarch – in clear and striking contrast to the French who addressed the citizenry and their representatives. Both, however, depart from traditional medical ethics, whose audience was meant to include physicians, perhaps only physicians. For Frank, as we shall see momentarily, this choice of audience has important consequences.

Frank characterizes his enterprise first in general terms and then in terms specific to the task of the medical police.

The internal security of the state is the subject of general police science. A very considerable part of this science is to apply certain principles for the health care of people living in society, and of those animals which they need for their work and sustenance ([14], p. 12).

He continues:

Medical police, like all police science, is an art of defense, a model of protection of people and their animal helpers against the deleterious consequences of dwelling together in large numbers, but especially of promoting their physical well-being so that people will succumb as late as possible to their eventual fate from the many physical illnesses to which they are subject ([12], p. 13).

From these statements and those of Rau we can set out the philosophical underpinnings of the concept of medical police. It is the monarch, the intended audience of these texts, who has the right to large, healthy and hardy working populations in both cities and countryside. The monarch, out of self interest in times of war as well as peace, has the need to make provision of the health of his or her citizens. Medicine, as an institution, thus is transformed into an agent of monarchal or state policy. On such a view, physicians do not owe their primary obligations to their patients, as the Hippocratic tradition would have it. Instead, they owe their duties *to* the state *regarding* the health of citizens.

Now, considerable authority thus devolves upon physicians as medical police, as agents of the monarch's rights and will. Frank was keenly aware of the implications of his theory of medical police in this respect: personal liberties might be jeopardized. In the Preface to the second edition (1783) of his work he asks:

Will an increased field of the police (as has been feared since the first publication of this work) thus immensely limit the natural freedom of man, which is being more and more curtailed anyway, abuse the rights of the family fathers, husbands, parents, and will that which is thus wrongly taken away from them be put into the despotic hands of the authorities? ([14], p. 10)

His initial answer is an interesting one.

... I shall gladly leave the decision to every philosophical observer of human nature. I only believe that I must add the following.

It is incomprehensible to me how anyone can hope to retain natural freedom in social life without curbs, and it seems to me this is philosophizing *a la* Rousseau. Can one not raise equally valid objections to all laws? ([14], p. 10).

But to raise such objections in the manner of a Rousseau would be to undermine the very possibility and foundation of the stable social order Frank has conceived. In any case, he goes on to say, the freedoms to be sacrificed are not important ones.

In all this and other tasks of medical Police, there is nothing that could possibly violate freedom in a community, or make the sensible citizen a slave of the law-giving authorities. They only take care of him and, so-to-say, take away from their children the knife with which they could injure themselves gravely ([14], p. 11).

The 'knife' in question Frank has earlier characterized as "a murderous trade with Aqua tofana, Poudres de Succession, and with abortifacients, etc." ([14], p. 11). In short, only the freedom to be reckless with one's responsibilities to the common good — preserving one's own and others' health — are lost and these limits on freedom reasonable people, certainly, should be willing to accept. Thus, significant freedom is preserved. At the same time, doctors have clearly defined for them their duties to the state regarding such widely diverse matters as the care of pregnant women, their fetuses, their babies, food, drink, clothing, housing, safety measures, and 'popular amusements.' Thus, the doctor's obligations are to go beyond the patient-physician relationship to embrace matters of the public weal.

The Virtuous Physician

Still another approach to matters of institutional medical ethics emerges from traditional, patient-centered 'Hippocratic' medical ethics. This approach has its roots in notions of the virtuous physician prevalent in the eighteenth and nineteenth centuries in Great Britain and America. The concept of the virtuous physician is not well understood by bioethicists, nor by students of the history of medicine and medical ethics. The typical view seems to be that virtue centers principally on matters of medical etiquette and has its roots in the notion of what a gentleman was, in particular the *Christian* gentleman

[7]. Both of these views, as I have suggested elsewhere, can be challenged and found wanting ([24, 25]).

It is important to remember that the university-educated physician in Great Britain and America in the eighteenth century was a man educated in the liberal tradition, in the best sense of that phrase. Then too, when we speak of the eighteenth century in the English-speaking world we are speaking of the Enlightenment, a time when deliberation about ethical matters was being quite self-consciously conducted on grounds independent of — and even antithetical to — religion. This is especially true of the Scottish moralists — Hume and Hutcheson particularly — whose influence on British and American physician-authors of treatises in medical ethics cannot be overstated. The chief of these are James Gregory (1753–1821) in Scotland and Samuel Bard (1742–1821) in America.

James Gregory held the chair in Physic at the University of Edinburgh, a position which his father, John Gregory (1729–1773) had occupied previous to his own tenure. Now, John Gregory was well known for his lectures on medical ethics, which were first published by his students [18] and later by their author himself [17]. Elsewhere [25], I have attempted to show the serious philosophical foundations of these lectures in the moral philosophy of David Hume. James Gregory's views on medical ethics appear to pick up where his father's left off, extending them still further.

James's views on medical ethics appear in his *Memorial to the Managers of the Edinburgh Infirmary*, published in 1800 [16]. This lengthy treatise is an exposé of what Gregory took to be abuses of patients used in experiments in the Infirmary. This treatise was not a private document but was published publicly and presumably meant for consumption both within and without the medical profession. Thus, like the French and German writers on matters of institutional medical ethics, the audience for Gregory's arguments was meant to embrace the broader public, including, as we shall see, the public officials responsible for the administration of the Infirmary.

Gregory's line of argument is rather straightforward: Physicians, by becoming physicians, have freely taken on certain obligations which then define them *as* physicians.

What I have particularly at heart is, to shew, that belonging to certain professions, or stations, or offices of trust, there are certain *duties*, which in their own nature, or from the nature of things are *supreme* and *indefeasible*; which no individual, and no set of men, can, either for themselves or their successors, violate, or renounce, or neglect, without substantial *injustice*; such injustice as law might prevent, or undo, or perhaps punish, but can never be supposed to sanction and enforce ([16], p. 134).

Notice already that Gregory is calling for public sanction of the physician's duties. With respect to human experimentation two conflicting duties emerge: the duty to improve a science that is beneficial to mankind and the duty to do what is best for one's patients while not exposing them to unnecessary danger. The latter, he argues, is among the "supreme and indefeasible" duties of the physician ([16], p. 144). That is, in determining the broad, social issues involved in the ethics of experimentation, the profession *and* the public must not establish duties that conflict with those owed to patients in the patient-physician relationship.

Gregory extends this argument in two striking ways. In the first, he argues for what amounts to the institutional rights of patients. What is to be respected are the following:

Even the *least* of the duties of a physician which are specified in our oath, and universally understood, independently of any oath, to be the *duty* of a physician, and consequently the *right* of his patients and families ([16], p. 136).

Note that the appeal is not a parochial one — simply to the doctor's oath — but is broadly based, "independently of any oath." In short, the obligation to do what is best for one's patients while not exposing them to unnecessary danger can be established without appeal only to what doctors and doctors alone could know [36]. Gregory is not clear on what these other grounds would be, but one suspects — and this has yet to be shown — that he had his father's medical ethics in mind, which had its roots in the ethics of sympathy of David Hume. In any case, the important feature is that Gregory argues that there are not simply duties of virtue involved but that these duties generate or confer rights on patients within an institutional setting. The duties in question Gregory describes as follows:

Whatever was best for his patients, it was his indispensible professional duty to do for them, whatever was bad, or unnecessarily dangerous for them, it was his duty not to do; and both of these duties were with him supreme and indefeasible ([16], p. 144).

We would say that the rights of patients in this context were not contractual but, as Siegler [33] and May [23] would have it, covenantal. But the rights of patients emerge, it would seem, only within (not independently of) the patient-physician relationship. In this institutional setting, one would argue for a right to health care which surely included the 'decent minimum' but was not limited easily by cost/benefit concerns.

The second interesting feature of Gregory's argument is his view on the provisions for enforcement of these professional duties and the rights of patients which those duties confer. Enforcement shall *not* be left to the medical profession, but becomes the responsibility of officials beholden not to the profession but to the public, through the monarch; namely, the managers of the Royal Infirmary.

Whatever it is the duty of Physicians and Surgeons to *do* to their patients, it is the duty of the Managers of an Hospital to *procure* for the sick from who are admitted to it.
Whatever it is the duty of Physician and Surgeons *not to do* to their patients, it is the duty of the Managers *not to permit* in their hospital ([16], p. 145).

In short, issues concerning duties of medical personnel and the rights of patients have now become public matters, a key feature of institutional medical ethics. While for us this may not be a remarkable development, in Gregory's day, in which physicians' and surgeons' societies held Royal Charters granting monopolies to practice, a public exposé of this sort was unthinkable.[1] To be sure, Gregory does not address such questions as the right to health care or resource allocation. But he does set out the grounds for doing so: the concept of the virtuous physician. An approach to institutional medical ethics of this kind would take its start in and seek always to be adequate to the basic ethical dimensions of the patient-physician relationship.

A similar line of reasoning was developed, several decades earlier and under another sort of philosophical influence, in America by Samuel Bard (1742–1821), one of the founders of King's College medical school in New York City. Bard was a native American who had been educated at the medical school in Edinburgh in the middle of the eighteenth century. He addressed, in a commencement essay at the new medical school in 1769 [2], matters that could be regarded as institutional medical ethics: provision of health care for the sick poor and the establishment of a public teaching hospital.

Bard's approach to these matters and to the ethical dimensions of the patient-physician relationship is the same: the application of Francis Hutcheson's moral sense theory [20]. This view held that we are all naturally disposed toward benevolence, other-regarding actions, and to approve such actions. Thus the very best example of right action, in Hutcheson's view, would be that which was other-regarding and never that which was (exclusively) self-regarding. Moreover, there is a clear, indeed paramount, obligation to perform benevolent actions. As Hutcheson puts it, rather forcefully in the context of his own philosophy: "there is naturally an obligation upon all men to benevolence" ([20], p. 249), and benevolent actions are those directed to the "publick good" ([20], p. 172).

Bard adopts Hutchesonian reasoning in this account of both the ethics of the patient-physician relationship and of the broader, social responsibilities of physicians. There is thus, for example, an obligation to treat the sick poor without remuneration. To refuse to do so, Bard argues, would be to act contrary to benevolence, to become a true moral wretch, deserving respect from no one.

The physician's obligation to the public good, though, does not stop at the individual level. They are obligated, according to Bard, to establish and contribute services and funds to a public hospital in which all of the sick poor of New York City would be treated. This obligation does not have its origins in basic rights, as it would later for the French, but in the concept of the virtuous physician, one who acted from duty and benevolence.

Much later, in the first Code of Ethics of the American Medical Association, promulgated in 1847, we can discern a similar line of thought, though one more seemingly devoid of explicit philosophical roots.

Of the Duties of the Profession to the Public, and of the Obligations of the Public to the Profession
Art. I — Duties of the Profession to the Public 1. As good citizens, it is the duty of physicians to be ever vigilant for the welfare of the community, and to bear their part in sustaining its institutions and burdens: they should also be ever ready to give counsel to the public in relation to matters especially appertaining to their profession, as on subjects of medical police, public hygiene, and legal medicine. It is their province to enlighten the public in regard to quarantine regulations, — the location, arrangement, and dietaries of hospitals, asylums, schools, prisons, and similar institutions, — in relation to the medical police of towns, as drainage, ventilation, etc., — and in regard to measures for the prevention of epidemic and contagious diseases; and when pestilence prevails, it is their duty to face the danger, and to continue their labours for the alleviation of the suffering, even at the jeopardy of their own lives [1].

The spirit, though, is similar to that of Bard and Gregory: physicians have obligations as a result of an understanding of what the conscientious or virtuous physician, the physician-citizen in this case, should be. No mention is made of the basic rights of fellow citizens nor of duties to the state.[2]

In summary, then, the British and the American views on institutional medical ethics, during the period in question, take their start in concepts of what a virtuous physician should be and what he, therefore, owed to his patients and the public at large. Though by the middle of the nineteenth century the philosophical foundations of this view are less clear, in the previous century there can be little doubt of the Enlightenment character of the thought of physicians like Gregory and Bard and of their debt to the Scottish moralists.

Their approach is strikingly different from that of the French, which takes its start external to both the patient-physician relationship and particular social orders. Instead, for the French, respect for and realization of basic, natural rights shall be the foundation for an ethic of institutional medicine. The Anglo-American approach similarly differs from the concept of medical police. On that view, clearly, the starting point is not the patient-physician relationship, but duties owed to the state regarding the character of that relationship. On such a view, matters concerning the ethical dimensions of the patient-physician relationship become secondary to matters of an institutional character, the reverse of the view based on a concept of the virtuous physician.

IV. THE BEARING OF HISTORICAL INQUIRY

If the preceding historical account of eighteenth and nineteenth century examples of institutional medical ethics is correct, if only in broad outline, then it becomes plain that there are rather ample historical precedents and models for contemporary inquiry into this complex set of concerns. I close with some suggestions about the bearing of these historical inquiries on our own efforts, with an emphasis on what was then and is now the central feature of that inquiry.

First, though, I want to outline how historical efforts complement our own. This is especially true of two of the approaches treated in the initial topography. Consider again, for example, the views of Childress [9] and Siegler [33]. The latter's appeal is to traditional views of the patient-physician relationship, which, we are told, is covenantal in character. Childress suggests development of this view and others like it along value rational lines: "the value rational approach focuses on the values that are expressed in policies" ([9], p. 44). We learn little, though, about just what values *should* be addressed in policies for health care.

Here, the inquiries of an earlier era into the concept of the virtuous physician have bearing, for they point to the way in which the argument must develop, *viz.*, in terms of duties to patients and to the public which are proper to the role of physician or, more broadly, the health care professional. For Gregory and especially for Bard the basis of those duties was understood in terms of how duties founded in the patient-physician relationship could be extended to embrace the public duties of health professionals, physicians in particular. Now, we may disagree with the method of analysis and argument they employed, that of the Scottish moralists, but we cannot disagree with their starting point: the freely assumed duties of health providers, first to their patients, and second to the broader public and the institutions that serve the public good.

Our task is much more difficult, however, than was either Gregory's or Bard's, and in at least two ways. First, as was pointed out earlier, the expectations for and demands on medicine have increased enormously since the eighteenth century. Medicine increasingly has the power to aid in our search for individual and common goods and we expect it to deliver. These increasing expectations and demands have helped to make medicine as a social enterprise more complex and difficult to manage: whose ends, needs, wants, or rights shall be served when not all can be? Second, we do not share widespread confidence in a single underlying ethical theory to ground and direct our own inquiries. To be sure, justice is at the center of our concern, at least in this country, but justice is a devilishly difficult concept to grasp and employ. Perhaps a Rawlsian maximin theory [29] will serve, perhaps not. When one places the issues in an international context the prospects for consensus on the basic presuppositions for the inquiry into justice and health care grow dimmer still. Thus, while the similarity to present concerns of previous efforts may aid us in augmenting, if not completing, our own inquiries, at the same time their differences from the present context bring into sharper focus the more complex and difficult character of our own undertaking.

I would like to close with a consideration of what I take to be the heart of that complexity and difficulty: the proper balance to be struck between the ethics of the patient-practitioner relationship, on the one hand, and institutional medical ethics, on the other. The tension between these two can be seen already in Frank's concept of medical police. Under a scheme in which the duties of health professionals are owed to the state regarding the health of the population, are not inevitable conflicts to occur between such duties and those owed to persons in the patient-practitioner relationship? The answer, it would seem, is 'yes'.

Consider, for example, Frank's treatment of the morality of abortion. Not only is such an act an "entirely unnatural crime" ([14], p. 103) because it is murder, abortion also — and primarily, it would seem — is injurious to the state, for it is "robbed" of a future citizen by abortion ([14], p. 104). That is, whatever duties one may have to the mother, obligations to the fetus and to the state *regarding* the fetus are to be pre-eminent, especially the latter. Now, on, let us say, a utilitarian account of institutional medical ethics, a similar sort of conflict could emerge. If one's primary obligation is to the maximization of the public good, then obligations to one's patients or clients would become secondary.

Thus, if one argues in the straightforwardly utilitarian manner of a Joseph Fletcher [11], let us say, then one arrives at a scheme that seems injurious to the needs, interests, wants, and rights of individuals who find themselves patients in the health care system. Against Hans Jonas's view that the physician is obligated to the patient and to no one else Fletcher claims:

In medicine we have to deal with many patients who coexist, for instance, not just with patients one at a time. Hans Jonas says, "In the course of treatment the physician is obligated to the patient and no one else." That, alas, is the classical ethics at its most myopic. It looks at the problem of medical care through a rear-view mirror ([11], p. 104).

Fletcher's glib, McLuhanesque cliché, however, does not serve as serious philosophical argument. What one wants is an account of the proper balance of the health professional's obligations, and Fletcher does not deliver. One might, alternatively, take a more modest, modified utilitarian approach, along the Rawlsian lines developed by Ronald Green [15]. On this more balanced view it is possible to give some place to the individual person who, after all, remains a patient, despite Fletcher's own myopic view of matters.

Still, even on a Rawlsian view of institutional medical ethics, there may be a problem. Roy Branson accurately characterizes the Rawlsian project in the following way:

If bioethics were to follow Rawls, it would not simply add the interinstitutional concerns of social ethics to its agenda; bioethics would make certain that considerations of social justice took precedence over analysis of individual obligations ([6], p. 14).

Clearly, on such a view, the potential for conflicting duties, of the kind we see in Frank's account of medical police, is very great indeed.

Albert Jonsen and André Hellegers, by contrast, appear to be of the view that no conflict would emerge in the attempt to provide an adequate theory of the justice of medicine as a social enterprise [21]. Others, however, are not so sanguine. Siegler is opposed to at least one approach, the rights to health care approach, because it would be in contradiction to what he claims is the traditional view, i.e., a non-contractual, covenantal model [35]. The roots of Siegler's view are in the Anglo-American medical ethics of the eighteenth and nineteenth centuries, though he does not seem aware of this point.

This tension between individual obligations and institutional justice is tied, I think, to the problem of identifying the intended audience of the debate. For Bard and the AMA's first code no strain, much less conflict, between the two emerged, it would seem, because the intended audience was restricted to physicians. Gregory's audience included more than just physicians, but only, it turns out, in the limited role of moral/legal enforcers of physicians' obligations to their patients, duties which for him were indefeasible. This conclusion is unremarkable, given the primary focus on the ethics of the patient-physician relationship.

When one, however, addresses a broader audience and thus the broader subject matter of institutional medical ethics, as the Germans and French did, the strain emerges. This is especially evident in Frank's work, in which, as we have seen, he gives evidence of concern for some of the implications of his system for personal freedom. Against the French, one might lodge Siegler's objections. Thus, contemporary approaches through theories of rights, utility, or justice to institutional medical ethics will, inevitably, it would seem, encounter conflicts between individual and social obligations of health professionals, when concern for social obligations becomes preeminent — as it must when the intended audience for accounts of institutional medical ethics expands to include the general public and their servants in government.

Contemporary arguments based on concepts of the virtuous physician may escape this conflict, but at a price. Those who lived in earlier times, even in the early Enlightenment, had an advantage over us — their times were considerably more homogeneous than our own. Thus, the task of articulating individual and social obligations in a "value rational" manner was conducted against a background of widely shared assumptions about medicine, physicians, and virtue (See [28], p. 34). In today's world this is not the case. We live in a pluralistic society and certainly a pluralistic world, and dilemmas relevant to institutional medical ethics must take this into account. Thus, resort to concepts of virtue and obligation will succeed only against a background of common values — and there are many such backgrounds within medicine and certainly in the lay public. An over-arching account of virtue and obligations would thus appear an impossible task.

Thus, as we attend more closely to the achievements of our forebears on this complex and difficult inquiry we may find that we will be able to achieve little more than they did: a series of very different accounts appropriate to the peculiarities of the historical contexts in which they emerged and to which they were addressed. In short, the task may be a quite modest one, that of attempting to say how best to strike a balance between the individual and social levels of ethics with the explicit recognition that such a balance will be temporary at best. From the point of view of public policy formation this may well be an altogether healthy development, for it will remind us that the best public policies, regarding matters as complex as justice and health care, must be open to criticism and revision. They will be so only if we recognize that they are as tentative, uncertain, incomplete, and temporary as were their historical forebears.

Georgetown University School of Medicine and
Kennedy Institute of Ethics
Washington, D.C.

NOTES

[1] The detritus of Gregory's exposé was sad. In a series of exchanges with his colleagues, totalling over 1500 published pages, Gregory alas resorted to ad hominem arguments, invective, and pettiness — as did his critics. As a consequence, his fine moral argumentation was not appreciated by his contemporaries and appears to have had little effect.

[2] In this respect, though the term 'medical police' does not appear, its meaning is not the same as it was for Frank. After all, the political philosophy of democracy is at a far remove, if not contradictory to, Frank's views on medical police. As Rosen ([30], pp. 154–155) points out, in the United States, as in Great Britain, the central meaning of medical police concerns provision for public health like safe housing, sanitary water supplies, proper sewage disposal, and the like and *not* the more systematic and wide-reaching German scheme.

BIBLIOGRAPHY

1. American Medical Association: 1847, 'First Code of Medical Ethics', *Proceedings of the National Medical Convention 1846–1847*, 83–106.
2. Bard, S.: 1769, *A Descourse upon the Duties of a Physician* . . . , A. & J. Robertson, New York.
3. Barnett-Connor, E.: 1979, 'Preventive Medicine and Public Health in the People's Republic of China', *Preventive Medicine* 8, 567–572.
4. Beauchamp, T. and Faden, R.: 1979, 'The Right to Health and the Right to Health Care', *Journal of Medicine and Philosophy* 4, 118–131.
5. Bell, N. K.: 1979, 'The Scarcity of Medical Resources: Are There Rights to Health Care?', *Journal of Medicine and Philosophy* 4, 158–169.
6. Branson, R.: 1976, 'The Scope of Bioethics: Individual and Social', in R. M. Veatch and R. Branson (eds.), *Ethics and Health Policy*, Ballinger Publishing Company, Cambridge, Massachusetts, pp. 5–16.
7. Burns, C. R.: 1978, 'North America: Seventeenth to Nineteenth Centuries', in section on 'Medical Ethics: History of' in W. T. Reich (ed.), *Encyclopedia of Bioethics*, Macmillan and Free Press, New York, pp. 963–968.
8. Callahan, D.: 1976, 'Biomedical Progress and the Limits of Human Health', in R. M. Veatch and R. Branson (eds.), *Ethics and Health Policy*, Ballinger Publishing Company, Cambridge, Massachusetts, pp. 156–165.
9. Childress, J.: 1979, 'A Right to Health Care?', *Journal of Medicine and Philosophy* 4, 132–147.

10. Engelhardt, H. T., Jr.: 1979, 'Rights to Health Care: A Critical Appraisal', *Journal of Medicine and Philosophy* 4, 113–117.
11. Fletcher, J.: 1976, 'Ethics and Health Care Delivery: Computers and Distributive Justice', in R. M. Veatch and R. Branson (eds.), *Ethics and Health Policy*, Ballinger Publishing Company, Cambridge, Massachusetts, pp. 99–110.
12. Fried, C.: 1976, 'Equality and Rights in Medical Care', *Hastings Center Report* 6, 30–32.
13. Frank, J. P.; 1768–1817, *System einer vollständigen medicinischen Polizey*, 3d. rev. ed., Wien.
14. Frank, J. P.: 1976, *A System of Complete Medical Police: Selections from Johann Peter Frank*, E. Lesky (ed.), Johns Hopkins University Press, Baltimore, Maryland.
15. Green, R. M.: 1976, 'Health Care and Justice in Contract Theory Perspective', in R. M. Veatch and R. Branson (eds.), *Ethics and Health Policy*, Ballinger Publishing Company, Cambridge, Massachusetts, pp. 111–126.
16. Gregory, J.: 1800, *Memorial to the Managers of the Edinburgh Infirmary*, Murray and Cochrane, Edinburgh.
17. Gregory, J.: 1772, *Lectures on the Duties and Qualifications of a Physician*, W. Strahan, London.
18. Gregory, J.: 1770, *Observations on the Duties and Offices of a Physician*, W. Strahan and T. Cadell, London.
19. Hippocrates. 1923: *Hippocrates*, W. J. S. Jones (trans.), The Loeb Classical Library, Harvard University Press, Cambridge, Massachusetts, Volume I.
20. Hutcheson, F.: 1971, *An Inquiry into the Original of our Ideas of Beauty and Virtue*, Georg Olms Verlag, Hildesheim (originally published in 1725).
21. Jonsen, A. R. and Hellegers, A. D.: 1976, 'Conceptual Foundations for an Ethics of Medical Care', in R. M. Veatch and R. Branson (eds.), *Ethics and Health Policy*, Ballinger Publishing Company, Cambridge, Massachusetts, pp. 17–34.
22. Locke, J.: 1960, *Two Treatises of Government*, P. Laslett (ed.), Cambridge University Press, Cambridge, England.
23. May, W. F.: 1975, 'Code and Covenant or Philanthropy and Contract', *Hastings Center Report* 5, 29–38.
24. McCullough, L.: 1978, 'Great Britain and the United States in the Eighteenth Century', in section 'Medical Ethics: History of' in W. T. Reich (ed.), *Encyclopedia of Bioethics*, Macmillan and Free Press, New York, pp. 957–963.
25. McCullough, L.: 1978, 'Historical Perspectives on the Ethical Dimensions of the Patient-Physician Relationship: the Medical Ethics of Dr. John Gregory', *Ethics in Science and Medicine* 5, 47–53.
26. McCullough, L.: 1979, 'The Right to Health Care', *Ethics in Science and Medicine* 6, 1–9.
27. Percival, T.: 1975, *Percival's Medical Ethics*, C. Leake (ed.), Robert E. Kreiger Publishing Company, Huntington, New York (originally published in 1803).
28. Ramsey, P.: 1976, 'Conceptual Foundations for an Ethics of Health Care: A Response', in R. M. Veatch and R. Branson (eds.), *Ethics and Health Policy*, Ballinger Publishing Company, Cambridge, Massachusetts, pp. 35–36.
29. Rawls, J.: 1971, *A Theory of Justice*, The Belknap Press of Harvard University Press, Cambridge, Massachusetts.

30. Rosen, G.: 1974, *From Medical Police to Social Medicine: Essays on the History of Health Care*, Science History Publications, New York.
31. Sade, R.: 1971, 'Medical Care as a Right: A Refutation', *New England Journal of Medicine* 285, 1288–1292.
32. Sidel, V. W. and Sidel, R.: 1976, 'Self-Reliance and the Collective Good: Medicine in China', in R. M. Veatch and R. Branson (eds.), *Ethics and Health Policy*, Ballinger Publishing Company, Cambridge, Massachusetts, pp. 57–75.
33. Siegler, M.: 1979, 'A Right to Health Care: Ambiguity, Professional Responsibility, and Patient Liberty', *Journal of Medicine and Philosophy* 4, 148–157.
34. Veatch, R. M.: 1978, 'Issues of Informed Consent in Human Experimentation', in L. B. McCullough and J. P. Morris (eds.), *Implications of History and Ethics to Medicine – Veterinary and Human*, Texas A & M University, College Station, Texas, pp. 90–106.
35. Veatch, R. M.: 1977, 'Just Social Institutions and the Right to Health Care', *Journal of Medicine and Philosophy* 4, 170–173.
36. Veatch, R. M.: 1972, 'Medical Ethics: Professional or Universal?', *Harvard Theological Review* 65, 531–559.
37. Veatch, R. M.: 1976, 'What Is a "Just" Health Care Delivery', in R. M. Veatch and R. Branson (eds.), *Ethics and Health Policy*, Ballinger Publishing Company, Cambridge, Massachusetts, pp. 127–153.
38. Wiener, D. B.: 1978, 'France in the Nineteenth Century', in section 'Medical Ethics: History of', in W. T. Reich (ed.), *Encyclopedia of Bioethics*, Vol. 3, Macmillan and Free Press, New York, pp. 971–973.

SECTION II

ISSUES OF MICRO-ALLOCATION

ERIC J. CASSELL

DO JUSTICE, LOVE MERCY:
THE INAPPROPRIATENESS OF THE CONCEPT OF JUSTICE
APPLIED TO BEDSIDE DECISIONS

Discussions about the allocation of scarce resources to individual patients have frequently been based on the concept of justice. I am going to argue that it is usually not appropriate to ground individual treatment decisions on ideas of justice. Compassion and mercy are the moral concepts better suited to the inequalities and individual differences of the sick.

Understanding justice in the context of medical care is made more difficult by several common confusions. First, medical care is primarily concerned with the care of the sick and not with health. Second, the origins of national health insurance programs (which are, of course, sickness insurance programs) have more to do with decreasing the costs to the state of poverty, disease, disability, and premature death than concern with the individual sick person. Third, attempts to apply principles of justice at the individual bedside falter for a number of reasons, both symbolic and factual. Fourth, invoking principles of justice in the setting of individual sickness denies the reality and potency of individual differences as well as denying the existence of fate.

Let me first address the fact that medical care is primarily about the care of the sick and not about health. While it is certainly true that pneumonia is an impediment to health, not having or recovering from pneumonia is an insufficient criterion for health. But, preventive medicine, as conventionally discussed and practiced, is concerned with the prevention of disease, not primarily with the promotion of health. The word 'health' is often used in the United States as a euphemism to conceal the presence of 'sickness', That is the case, much as the Hospital for Special Surgery was originally called the Hospital for the Ruptured and Crippled. Sterility clinics became fertility clinics and contraceptive clinics became family planning clinics. Another example is a program being funded by a prominent foundation for the 'health impaired elderly'. Those words are a euphemism for the old and the sick (which is what they call themselves). Euphemisms are fine until they begin to fool their users. The euphemism works because most people put health and disease at the opposite end of the same spectrum. That understanding, however popular, will probably not withstand critical examination — especially in view of the difficulties of finding a definition of health. Whether or not health and disease are more than passingly related to each other, it *is* the case that medical care is overwhelmingly the care of the sick. The distinction is important because it seems easier to decide whether, in a just society, a sick person should receive care regardless of the ability to pay, than to decide whether a person has a right to health — whatever that may be.

Now for the second confusion. At first glance, it would seem that health insurance programs, or national health care systems, are really an extension of human activity that stretches back into antiquity — charity to the sick. What seems new in this century is the idea that all sick persons should have an equal right to medical care independent of their ability to pay. In that context, the word 'charity' is often considered onerous. It may be pertinent, however, to remember that national health insurance programs have their historical basis not primarily out of concern for the individual sick person but, rather, out of concern for the burdens of the state. Edwin Chadwick, who had been secretary to Jeremy Bentham ([6], pp. 32 ff), published his 'Report on the Labouring Classes of Great Britain', in 1842 [4]. The investigations that formed the basis for that publication grew out of ferment for reform of the 'poor laws' of Great Britain. Chadwick showed systematically how much more disease, disability, and premature death were to be found among the poor than among the comfortable. Whatever his humanitarian concerns, his basic argument for removing that disease burden by sanitary reform was that the poor were poor (and a drain upon the state) because they were sick (Chadwick's observation that the poor have more sickness than the comfortable has been repeated generation after generation and remains true today in the United States and even in Great Britain). What Chadwick proposed was pure Bentham utilitarianism. The legislation that created Britain's National Health Service was a direct descendent of the poor law reform of the mid-19th Century. In view of the history of National Health Insurance, its fundamental objectives, and its philosophical foundations, questions of distributive justice are entirely appropriate, as are discussions of a right to medical care. But the focus of the concern remains, not the care of a particular sick person, but that person's needs in relationship to the needs of others and in the light of the resources of the state.

But the debate about justice and medical care has gone further: Now we ask what are the limits of the equal distribution of medical services (regardless of ability to pay) and, more, how are scarce resources to be allocated? This is often symbolically phrased: Which of several patients with end-stage renal disease should get 'the kidney'? At first glance, the sick-room appears to be an appropriate place to examine the concept of justice. The sick seem an apt reminder of why the concept must first have arisen. The idea of justice is a response to the plight of persons subject to a power over which they have little control and injured by that power beyond anything they deserve. They have not been rendered to as was their due (if I may draw on one of the earliest definitions of justice).

In the beginning, and in most discussions, the concept 'justice' is employed in opposition to the concept of 'injustice'. If that is the case, justice may not be the appropriate concept when illness is considered because, when those terms are used, human agency and interest is involved (even in the derivative words applied to God, the community or the state). However, the sick are not sick because of human agency and intent but overwhelmingly because of the action of fate. Fate is called fate precisely because it is without human intent – fate cannot be unjust nor can it be just. The category of justice is simply not relevant. It has become fashionable to displace fate by speaking as though the sick have largely made themselves ill by their ways of life, that illness comes because of someone fouling the environment. But both ideas are naive because, ultimately, everybody sickens and dies no matter what their life style and no matter how clean the environment. Even habits which are known to be illness-producing, such as cigarette-smoking, produce sickness in only a minority of their habitués. Even when sick persons do things which seem (inexplicably) calculated to make themselves worse, it is almost never because the sick actively wish to be sicker. Such behavior merely testifies to the complexity of the state of illness and of the human condition.

An example, albeit over-simplified, of a situation often discussed which employs the concept of justice might be helpful: Two sick persons lie in adjacent beds – one is poor and one is rich, but both are in severe pain. As the doctor goes by, he or she gives pain relief only to the patient who can pay. Would we characterize the doctor as unjust, or would we say he is cruel, without mercy, has no pity, or lacks all compassion? Another example: two persons are dying of diseases similar in absolutely all respects. There is medicine (or a kidney, a treatment, a machine) to save only one. Who shall get the medicine? The question has been repeatedly argued on the basis of principles of justice. On that basis, I find no answer satisfactory. If the younger, more productive, smarter (use what criteria you wish) receives the treatment, you may argue the justice of the solution because he or she was due more. But the other patient died – was that also his or her due? Even by lottery, justice is not served because one must die. Indeed, we resort to lottery where no just solution appears possible. One might say that the lottery is the only fair way out of the difficult situation. Despite Rawls's usage, not all that is fair is just and not all applications of justice are fair. They are not equivalent concepts. It seems to me that none of these decisions need introduce the language of justice. Justice is not the relevant criteria, for no other reason than there can be no just solution — a solution in which each person receives his or her due.

Perhaps this is where the actions of a state and an individual are quite disparate. Justice may be the applicable term when the state is faced with the treatment of its sick citizens and where resources are limited. However, public policy and individual actions may differ in their moral basis, although apparently sharing the same goals.

Applying the concept of justice to individual decisions made at the bedside has additional difficulties. Addressing the question of which patient shall get the life-saving treatment (where there is only enough for one), some have suggested that the person who has in the past or can be expected in the future to contribute more to society, has the greatest claim to be saved. The paradox often present below the surface in discussions of justice at the bedside is revealed to some degree by that belief. There, the patient is being considered as a means towards the ends of the community or state, something which, at least in other circumstances, may be considered wrong. From the state's point of view, it is perfectly reasonable to save the person who will contribute the most to the state. Others believe that personhood is, in itself, sufficient to make all patients equal in that situation of scarcity. Those who hold this latter position want this fateful decision to be made solely on relevant 'medical' criteria. Who the patient was, is, or will be is not, for them, relevant. But, perhaps, people who want such awful decisions made solely on 'medical' grounds would also like to see patients treated as persons. And they are correct in that desire because we have all become aware of the problem of an overly technical medicine — physicians who care more about diseases and technology than about patients. Treating patients as persons is not only ethically warranted, but better medical practice. That is because the nature of the person cannot be separated from diagnosis, treatment, or prognosis, except artificially or in a trivial sense ([1, 3]). The obvious paradox is not easily resolved. The symbol of justice is a blindfolded woman who holds a technical device, the scales, with which to weigh all the relevant details without regard to the person of the disputants who stand before her. That has not been, at least until quite recently, an appropriate symbol for medicine. Indeed, the concept that applies in the symbol of justice is the political concept — equal before the law.

But in no other respect except the fact of their personhood are people equal. Here, equal is no longer a political term but applies to the biological, psychological or social factors involved in sickness. Some persons are taller, some smaller, heavier, smarter, with more or less hair, light-skinned and dark, strong or weak, and with personalities and dispositions that differ quite considerably. These differences are observable at the subcellular level as well as in the behavior of the family. The methods of modern science have been devised to reduce the impact of individual differences so that generalizations become possible upon which scientific advance can be made. But the fundamental fact of individuals is not sameness but particularity. Kenneth Schaffner [8] has pointed out how necessary it is for medicine to begin to understand how to go from the general to the particular. Not merely because physicians treat individual patients — in the full sense of the word individual [5] — but because those individual differences from the subcellular level to the family have a profound effect on the diagnosis, treatment and outcome of an illness.

The fact is that in the political system, as in medicine, we just do not know what to do with problems caused by individual differences. Consequently, we may pretend that they do not exist or that they are the result of some human wrong which, when made right, will erase the differences. It is in the nature of Nature that most human differences result from the operation of chance. We have become so successful in overcoming the handicaps imposed on people by fate (driven by our fundamental political belief in equality) that we have come to believe not only that all persons are equal in all respects, but that fate is fiction. Much current malpractice litigation seems primarily intended to compensate patients whose illnesses had an unexpected adverse outcome. This view assumes, falsely, that if something goes wrong, *somebody* must be at fault. There, a system of justice intended to protect people from being wronged by other people is used to indemnify them against the action of fate — a non-human agency.

What, then, is to be the moral basis of behavior toward the sick — especially when painful decisions must be made? Love of humanity, compassion, and mercy, *not* justice, are the appropriate concepts to guide actions at the bedside. The logo of the New York Hospital is not blindfolded justice, but the Good Samaritan. The obvious truth of what I have suggested has become obscured. (That otherwise superb work, the *Encyclopedia of Bioethics* [7], has a number of entries devoted to justice, but there are *none* about compassion or mercy, I hasten to point out.)

The process by which we in the Western world have forgotten that the sick among us are cared for out of love of humanity, compassion, and mercy, is duplicated in the education of medical students. Applicants to medical school learn, even before their interview, not to say that they want to be doctors because they like to help people, although that is often the case, because it sounds too sentimental. They are given an intense (and necessary) scientific education that ultimately provides them with an incredibly complex set of recipes for the treatment of disease and an equally complex and effective system for deciding what recipe to apply in a given instance. By the time seven or eight years have passed, they may not even remember, not to mention publicly acknowledge, why they chose medicine in the first place. In recent years, with the recognition that medicine is a moral (or moral-technical) profession [2], courses and discussions dealing with ethical problems have been included in the medical school curriculum [9], but the degree of intellectual power and systematic training given to the moral side of medicine nowhere approaches what is given to the technical.

Justice is defined in the *Oxford English Dictionary* as "the quality of being just or righteous; the principle of just dealing; the exhibition of this quality or principle in action." Mercy is " the forebearance and compassion shown by one person to another who is in his power and who has no claim to receive kindness; kind and compassionate treatment in a case where severity is to be expected." (Definition 7 describes mercy as "acts of compassion towards suffering fellow creatures.") Compassion is "the feeling or emotion that comes from being moved by the suffering or distress of another and the desire to remove it, pity that inclines one to spare or to succor." Note the difference in the use of terms expressing feeling in these definitions. Such words are virtually absent in definitions of justice.

Compassion and mercy are old-fashioned notions. But their hold on us has not been diminished solely by the advance of science. Those words smack of the benefaction of the strong towards the weak. There have been times when the pious application of these ideas has excused the worst oppression of the poor and downtrodden. Perhaps the paternalism inherent in the corruption of those virtues had to be overcome in order to achieve the political freedoms we now enjoy. Indeed, many things merely paternal, compassionate, or merciful, extended towards the sick by care-givers, are still called patern*alistic* — as though those virtues cannot be exercised honestly without the intent to oppress, demean, or coerce.

It is strange how little we know about these virtues. They are related to interactions between persons in situations of unhappiness, pain, suffering, and tragedy — the condition of being sick. They have to do with aspects of personhood that are neither political nor psychological but based, instead, on the moral claims members of the human community have on one another. In this era more is known about the political and psychological than the moral. But how are those virtues to be applied to the allocation of scarce resources to individual patients? And how will they be used to decide who shall survive? Their application is the stuff of wisdom, and wisdom, we are told, cannot be taught. But surely wisdom is learned more quickly where it is valued. It will be most valued when we recognize that compassion and mercy must guide us in difficult decisions. Knowing that, we must begin to give them the thought and analysis that has been given to justice in the last few centuries. I am suggesting two quite distinct things. First, moral awareness does not exist in all persons at all times. In many individuals, not only physicians, the compassion necessary for the care of the very sick is simply atrophied through disuse. Moral awareness must be specifically awakened, shown to be appropriate to the physician's act, and then made manifest. Only then can moral awareness take its place alongside scientific training and clinical experience. It is often said that people (specifically physicians) cannot be taught to be compassionate – they either are or are not. Where is the evidence for that claim? Who has tried and failed and how did they try? Which leads to the second suggestion. We must study compassion in order to understand it and its applications if it is to be effectively taught. If that sounds excessively mechanistic, it is only the current habit of thought that makes it so. We did not arrive at the present sophisticated understanding of justice merely by having lived, but rather because of the active reflection on the concept, and its continual refinement. The same possibility exists for concepts like compassion and mercy. They are fruitful areas for study. It might sound odd to ask a student: 'On what basis would you extend or deny compassion to this patient?' But it is not so odd (or dangerous) as having somebody extend or deny compassion without giving it any thought. We would not wish to be judged solely on the untrained intuitions of those who must judge.

One final point remains to be discussed. For all its faults, no society has ever, in the world's history, been as just as ours. Equality before the law is fundamental to the principle of justice. Justice is, therefore, the moral precept that acknowledges the fundamental equality of all persons. But compassion and mercy arise out of human need. They are moral qualities that acknowledge fundamental differences between persons. Recognition of inequality before nature, sickness, and fate is basic to a compassionate society. Perhaps, in the absence of a just society, compassion could be conceived as an instrument of oppression. But justice without compassion, without recognition of the fundamental differences between persons, is also oppressive.

In this essay, I have attempted to show that justice is not the most appropriate concept for guiding actions at the bedside of the sick. Compassion and mercy are far more appropriate. The issue is not simply one of realizing the necessity for change, but one of recognizing that compassion and mercy, like other virtues, must not only be awakened in but also taught to future physicians.

Cornell University Medical College
New York, New York

BIBLIOGRAPHY

1. Cassell, E. J.: 1971, 'Nineteenth and Twentieth Century Environmental Movements', *Archives of Environmental Health* 22, 35–40.
2. Cassell, E. J.: 1973, 'Making and Escaping Moral Decisions', *Hastings Center Studies* 1, 53–62.
3. Cassell, E. J.: 1979, 'The Uses of Subjective Information in Clinical Patient Care', presentation at the Conference on Changing Values in Medicine, New York, November.
4. Chadwick, E.: 1965, *Report on the Sanitary Conditions of the Labouring Population of Great Britain (1842)*, M. W. Flinn (ed.), Edinburgh University Press, Edinburgh.
5. Gorovitz, S. and A. MacIntyre: 1975, 'Towards a Theory of Medical Fallability', *Hastings Center Report* 5, 13–23.
6. Marston, M.: 1925, *Sir Edwin Chadwick*, Roadmaker Series, Leonard Parson Ltd., London.
7. Reich, W., (ed.): 1978, *Encyclopedia of Bioethics*, Macmillan Co., Free Press, New York.
8. Schaffner, K.: 1979, 'On the Nature and Limits of Clinical Medicine', Presentation at the Conference on Changing Values in Medicine, New York, November.
9. Veatch, R. M.: 1977, 'Medicine, Biology and Ethics', *Hastings Center Report* 7 (Special Supplement, December), 2–3.

MARC LAPPÉ

JUSTICE AND PRENATAL LIFE

HISTORICAL PERSPECTIVE

It has only recently become possible to consider issues of justice for prenatal life. Of course, spiritual considerations of the sanctity of life, particularly within the Roman Catholic tradition, have long reinforced a high regard for prenatal life. However, neither the social conditions which could have encouraged a view of fetal life as worthy of protection nor the medical breakthroughs which could have provided detailed knowledge of the exigencies of prenatal development were present before the mid-twentieth century. While many previous cultural groups reinforced familial cohesiveness, maternal care and, thus, indirectly, support for prenatal life, knowledgeable decisions regarding the medical and biological needs of the fetus are a modern day phenomenon.

Life in pre-industrial societies was generally so fraught with fetal and infant death, that social customs evolved which formalized neglect of the fetus and the newborn. Eskimo and Hawaiian societies, for instance, would only permit a child to be named *after* it had reached its first birthday, and the attainment of the fifth year (in the Hawaiian Islands) was cause for great celebration — because only then was there sufficient expectation of survival to permit formal acknowledgement of the infant's standing in society. Minimal investment in the newborn meant, of course, minimal investment in prenatal care. Prospective mothers were like as not left on their own, with little additional support or societal recognition. In most preliterate societies the mother was expected to continue her social duties during and, to a lesser extent, after pregnancy.

The Renaissance was an exception. Beatification of the infant Jesus and increased recognition of the fetus and newborn as an art object[1] probably reflected a renewed societal optimism. But social conditions at the outset of the Industrial Revolution, rather than improving, actually deteriorated with respect to the standing of the mother to be. The exigencies of the labor-intensive life marred the societal commitment to the child and her predecessor, the fetus in utero. Child labor and almost universal injustice in the treatment of youth were the hallmarks of the Industrial Revolution both on the Continent and in England. These abuses and a continuing high birth and infant morality rate climaxed in a paroxysm of disregard for the welfare of the newborn and its prenatal existence.

In the early 1600's, it was common to find children as young as five working the mines and mills of Britain and France. In 17th century London idle and needy children who wandered freely in the streets could be brought to the courts for meting out punishment. Hogarth in his famous engraving, *Gin Lane* (circa 1750's), showed prospective mothers swilling gin and neglecting their still nursing infants, while the rest of society merely ignored their plight, or worse, encouraged their husbands to put them to work in the mills alongside their other children. Women in both Britain and the New World were averaging eight children during their child-bearing years, only 1/3 to 1/2 of whom could be expected to survive. At the height of the gin-drinking era in England between 1740 and 1742, burials in London outnumbered baptisms two to one! And in England generally, children who were orphaned or abandoned outright might as readily be sent to the Colonies to be indentured servants, as be physically punished under the law. By the early 1800's the courts systematically treated offending minors as adults, regarding suspects as young as six as competent persons if some ingenious public defender could demonstrate that they knew the difference between right and wrong [6].

SOCIAL AND MEDICAL TRENDS

In such a world it was no accident that prenatal life received little or no attention. The two keys to the awakening of an interest in prenatal life and the conditions surrounding its nurture were the fall in the disastrously high rate of infant mortality at the outset of the 20th century, and the concomitant growth in social and medical attention to the plight of the poor. Most critically, medical research at the turn of the century began forcefully to demonstrate that the quality of life enjoyed by adult members of society is heavily determined by the quality of existence afforded to the fetus in utero.

By the late 1920's, studies of the results of the Irish Potato Famine were beginning to show just how far deprivations of prenatal life could affect the quality of life of the next generation. The retardation of pelvic bone growth which resulted from intrauterine starvation could lead to a second generation of deprivation as the frail bodies of undernourished girls undertook to carry their own pregnancies. Most often, the outcome of such post-famine pregnancies (particularly in Scotland) were small-for-gestational-age infants, or a mother, dead in childbirth from a too-small pelvic diameter. More contemporary studies give credence to the 'evil wreaked on generations' by intrauterine starvation by demonstrating how experimental animals whose own mothers were nutritionally deprived went on to produce a second generation of small brain size/small brain cell number offspring, even under the most ideal conditions of their rearing [11].

Public health trends also focused attention to the prenatal period. As the patterns of infant mortality shifted downwards with the advent of modern concepts of hygiene (elimination of puerperal sepsis was brought about by the simple expedient of surgeons washing their hands *before*, not after delivery), and later through the judicious use of antibiotics, the causes of perinatal death moved away from avoidable factors in the post-partum period to seemingly unavoidable ones inter-partum. Diseases like gastroenteritis, measles and respiratory infections gave way to congenital malformations as the major cause of death in the perinatal period by about 1960 in the United States. Such statistics point out a special relevance of the state of public health vis à vis care for prenatal life. Countries whose vital statistics are still in the pre-industrial pattern of high perinatal mortality generally vest less in supporting fetal life through affording prenatal care and nutrition than do emerging countries (e.g., Mexico) that have escaped from this pattern.

Campaigns to focus on the rights of newborns to a healthy environment, pioneered by the National Foundation/March of Dimes in this country, have taken root in Mexico, and now are commonplace among the most effective public health programs of federal agencies and state departments of health in the United States. The federal government's Food and Consumer Services division (under the Department of Agriculture), for example, has instituted a Women, Infants, and Childrens Supplemental Feeding Program (WIC) which has served more than 1.3 million participants (through November, 1978) with dramatic improvements in their hemoglobin and hematocrit values and outcomes of pregnancy [4]. States like California, New Jersey, Pennsylvania, and Michigan have all begun a series of initiatives to offset trends towards higher infant mortality in medically underserved areas.

JUSTICE AND POLICY FORMULATION

Each of these commitments of public funds is informed by an often unspoken but nevertheless powerful commitment to justice for prenatal life. There is a sense in many of these programs that somehow fetal life deserves special investments of taxpayer funds. In California, the allotment of scarce public resources for prenatal care initiatives has in fact been predicated on a rudimentary analysis of the issues of justice which would be served by an appropriately formulated policy. Distributional justice was considered in weighing where and to whom funds would be allocated. Retributive justice permitted us to single out the areas which had previously received the most minimal health resources. And arguments of social justice were used to allocate scarce resources to the groups which had experienced the worst perinatal outcomes over the previous few years. The rudimentary applications of justice, though only skeletally worked through, did in fact constitute the manner in which at least one such program was structured — the Governor's prenatal care initiative in Oakland, California.

In East Oakland, appreciably more newborns are brought to term without prenatal care than elsewhere in the state. Such infants face the prospect of infant death as a result of prematurity, respiratory disease and hyaline membrane disease fully twice as often as do mothers who receive prenatal care in months 1–4. Preventable causes of death like these account for almost half of all fatalities among the newly born in Oakland. Sudden infant death syndrome accounts for another half of deaths in the remaining 11 months of the first year of life. All of these causes of death occur substantially (i.e., approximately 200 percent) more often among unmarried mothers, blacks, Latinos and the poor than among white or Asian Americans. In this case, policy advisers for the Department of Health Services used the arguments of justice to establish at least one pilot prenatal care program.

But the quality of prenatal life is also a societal concern because of the social and economic consequences of neglecting it. Deprivation of care leads to high risk babies, while access to prenatal care, adequate nutrition and social support systems promises more optimal outcomes of pregnancy. Where such recourse has been given, as in the California State Department of Health Services initiative in Oakland, dramatic reversals of these undesirable outcomes of pregnancy have been achieved. For instance, by the end of the first year of operation, a 15 percent drop in infant mortality in the high-risk areas of Oakland.

Justice can also be predicated on measurably improving the quality of life of newborns by attending to prenatal care among high risk mothers. Reducing the incidence of low birth weight, a prognosticator of poor developmental milestones later in life, is another meaningful approach to improving outcomes for underserved populations.

Such policy decisions were predicated on viewing fetuses as a class, and then only in the context of the fetal maternal relationship. Each individual fetus has no past; no social standing equivalent to an adult; and no true legal standing on which to compute societal obligations. In California's experience it was possible to grant fetuses special attention by computing the deservedness of *groups* of persons — here the black and Latino populations of an inner city — for a massive infusion of state funds. But what about the actual standing of the fetus itself? Is there some way to compute its worth, social desert, or rights so as to generate a true theory of justice for prenatal life?

THE STANDING OF THE FETUS

Any discussion of prenatal justice must begin by grappling with the standing of the fetus in the wake of the Supreme Court decisions on abortion, in particular, the seminal *Roe v. Wade*. An investiture of the fetus with rights or standing must survive the realization that abortion is now an almost universally available maternal 'right'. Until recently,[2] abortion could be freely chosen by almost any woman who elected it. You might well ask how the fetus is to be vested with rights, if fetal life can be so abrogated, seemingly at will, by the choice of a prospective mother to terminate it. Prenatal life is indeed vested with fewer legal rights than is postnatal life. The most vociferous critics of American abortion policy point to this relaxation of legal protection for prenatal life as an abandonment of the fetus to the whim or caprice of any woman's wishes.

Such a position is based on a serious misreading of the Court's intent in *Roe v. Wade*. In deciding Roe, the Court emphasized that it was giving primacy to the right of the mother to elect an abortion without state interference, only up to a point when it believed the fetus came under the full aegis of Court protection. The Court only sanctioned an unrestricted right to elect an abortion by a woman and her physician in the first trimester. In the second, it reserved the right to allow states to determine the medical conditions under which abortions could be done. In the third trimester, the Court emphasized that its interest in the fetus became compelling, particularly once the fetus had traversed that indefinable moment of capability of separate life. The Court thus carried forward the concept of a fetus as an emergent person.[3]

In the eyes of the Court, the fetus was an entity whose rights accrued incrementally as it acquired greater and greater human potentiality. This viewpoint projects a kind of romantic notion of the fetus as an unfolding of human potentialities. But the actual biology of fetal development underscores a very different view. And it is this biological perspective which should dictate social policy in the area of justice for prenatal life.

BIOLOGICAL CONSIDERATIONS

The historical view of justice reinforced by the ruling of the Burger Court specifies that the degree and kind of social justice granted human life must follow a gradient of developmental ontogeny marked by the stages of embryonic and fetal growth — and later by comparable stages of moral development. Such an outlook is flawed because it ignores the extent to which the roots of later flowering of the person are grounded in discrete periods in prenatal life. Contemporary studies of human development reveal how much damage can be done to post-natal life if prenatal life is subjected to environmental insults [10]. The fetus is exquisitely sensitive to long-lasting injury at critical stages early in its development.

The image of the fetus as an *homunculus*, a man in miniature, that simply grew within the confines of the uterus gave way in the early 18th century to the more modern view of the embryo as an organism capable of progressive development. This ontogenetic view of the fetus has dominated modern thought about fetal rights. The Court's notion that the fetus is not a person in the 'whole sense' is an artifact of the belief that humanhood is acquired incrementally.

There is little dispute that the characteristics necessary for the emergence of a whole person are to be found in the developing fetus, both in terms of very specific and time-bound gene-environment interactions, and in the overall developmental program. But in order for later development to occur, extraordinarily sensitive events must transpire early in pregnancy. For instance, for full development of the eyes and neuromuscular system to occur, an incredibly complex web of specific events must occur, beginning with the migration of cells from the neural crest (day 18) and culminating with the specific innervation of appropriate neuromuscular target organs (days 62–70). Connections between the retina and the optical cortex of the brain are similarly predicated on both genetically programmed and environmentally influenced events.

The most minor perturbation in this developmental program, particularly during the critical period of organogenesis (days 16–44), can fatally disrupt the developmental plan and cause birth defects or fetal death. The fetus is thus both exquisitely genetically programmed and exquisitely sensitive to its environment during gestation, and requires affirmative care and protection to ensure that the great majority can proceed through pregnancy with the prospect of normal development post-natally. This viewpoint rests on the value-laden premises that issues of justice must regard fetuses collectively and

JUSTICE AND PRENATAL LIFE 89

invest heavily in the desirability of 'normalcy' in the outcome of pregnancy. For instance, the logic of prenatal care programs is based on the correlation between early attention to the mother (months 1—4) and achieving 'normal' birth weights. The universally acknowledged adverse relationship between low birth weight and subsequent outcome of postnatal life dictates policies which protect fetuses — without regard to their known genetic differences.

What, then, about providing societal investments to permit women who are otherwise infertile to conceive a pregnancy? The now classic case of Louise Brown shows that the medical community can generally downgrade the possible adverse consequences of in vitro fertilization on the quality of fetal life and its outcome in favor of parental 'rights'. The broader experience with use of related fertility enhancing or substituting treatments provides further evidence for founding medical policy on rights other than those of the fetus.

DISTRIBUTIVE JUSTICE

The distribution and availability then of at least three different modalities affecting prenatal life deserve renewed scrutiny. Access to fertility clinics or treatments, prenatal care and nutrition, and prenatal diagnosis all are presently unevenly distributed throughout the country with varying consequences to the at-risk population.

Lower availability of fertility treatments for black women, who have twice the infertility rate as do white women, may have dramatic effects on family structure, cohesiveness and patterns of size. Similarly, fewer prenatal visits and the access which such figures imply can significantly reduce the prospects for safe, uncomplicated pregnancies — and the adequacy of fetal outcome and development of those that reach term. Limiting access to prenatal diagnosis, especially as it expands to encompass sex choice information, holds the potential of selectively denying a pregnancy-planning option for poor families.

In California, the first sets of previously unpublished data reveal a pattern which has been found elsewhere. The availability and provision of prenatal diagnosis is preferentially given to relatively wealthy, white and Japanese-American women in favor of those who may be poor or non-white ([2, 4]). As is shown in Table I, this pattern of distribution suggests what medical practitioners have suspected about other forms of care afforded to prenatal life: where new technologies are offered, they are used preferentially by those with the access to care and wherewithal to pay.

TABLE I

Amniocentesis users: gross combined annual family income June 1976–October 1979

[*California median income in 1977 $20,858**]

Year	Less than $6,000	$6,000– $12,999	$13,000– $18,999	$19,000– $25,999	$26,000– $32,999	$33,000– $39,999	$40,000– or more	
1976	35	107	110	150	67	23	62	
1977	104	385	462	551	284	144	330	
1978	157	493	572	861	483	250	606	
1979	198	317	328	458	313	278	448	
Total†	494	1303	1472	2020	1147	695	1446	8,577
Percent	5.8%	15.3%	17.1%	23.5%	13.3%	8.0%	16.9%	

* Bureau of the Census.
† Excluding 4,722 who did not report income.

The income levels of amniocentesis users follows a biphasic pattern, with one of every six women having a family income in excess of $40,000 and the remaining users corresponding closer to the pattern of income in the state as a whole. While prenatal diagnosis should be sharply distinguished from direct, supportive care in the sense that some 2.5 percent of such examinations lead to a positive diagnosis and likely abortion, the opportunity to avail oneself of this new knowledge source and the concomitant medical attention represents a social good which is presently also inequitably distributed with regard to race (see Table II and III for the most current figures on amniocentesis evaluation in California). Significantly more women of Japanese, Filipino or Caucasian ancestry use amniocentesis than do Latin American women (Table III). White women predominate in the user picture proportional to their numbers in the 35 year and older group.

Considerations of distributive justice also are relevant to whether or not prospective mothers should be given societal support in conceiving a pregnancy. While it will not prove possible for some time to project the consequences of in vitro fertilization on the quality of fetal life and its outcome, it is possible to review the experience with use of related fertility enhancing or substituting treatments.

Recent studies have shown that pretreatment of women with clomiphene citrate, an ovulation stimulating drug and, more dramatically, with prednisone,

TABLE II
Proportion of California women 35 years or older receiving amniocentesis in 1977

Live births	Total	White (incl. Sp. spking)	Black	Indian	Chinese	Japanese	Filipino	Other non-white
Total	347576	296133	31535	1695	3726	2464	6155	5462
No. born to women $\geq$ 35 yrs	17197	11343	1067	61	308	240	758	388
Percent $\geq$ 35 yrs	–	4.8%	3.4%	3.6%	8.3%	9.7%	12.3%	7.1%
Number receiving amnios	3449	2483	126	8	62	88	99	65
Number of amnios as percent of $\geq$ 35 yr population	20%	17.3%	11.8%	13.1%	20.1%	36.7%	13.1%	16.8%
Expected number amnios; significance	3449	2876.5 $p < 0.001$	213.99 $p \geq 0.05$	12.23 n.s.	61.77 n.s.	48.13 $p < 0.001$	152.02 $p < 0.01$	77.81 n.s.

TABLE III
Relative utilization of amniocentesis services by Latin American and non-Latin American women† in California

	No. Using P.N.D.*	No. Not Using P.N.D.	Total
White Spanish speaking‡ 35 years	315	5633	5948
White non-Spanish speaking 35 years	2483	5912	8395

$x^2 = 1305.6$.
$p < 0.0001$.
* P.N.D. = Prenatal diagnosis.
† Self-definition of amniocentesis users.
‡ Using white Spanish speaking population underestimates total number of Latin American women by no more than 5–10 percent (black Spanish speaking over 35 years 530, i.e., approximately ½ all blacks 35 years).

a hormone which permits the carrying of pregnancies in otherwise compromised women, can have adverse effects on fetal outcome [9]. Reporting in *Science* magazine, a group of researchers at Rutgers, the State University of New York at Stonybrook, and the Southern California Fertility Institute demonstrated that the intrauterine growth of fetuses in prednisone treated mothers was substantially retarded. While the results of these pivotal studies have since been questioned ([5, 8, 9]), they point out a little known fact: substantial medical resources and attention are currently invested in initiating and sustaining pregnancy in otherwise infertile women. To the extent that these endeavors generate increased costs and diminish the resources which may be devoted to equally deserving women whose pregnancies are in jeopardy because of social conditions, this and related investments in fertility enhancement raise serious issues of both social and distributive justice.

RETRIBUTIVE JUSTICE

Failure to regard the prenatal period as one of exquisite vulnerability may lead to episodes of damage to prenatal life. The most dramatic such incident was the thalidomide disaster of 1960–1962 in which over 8000 children were born world-wide with substantial birth injuries that crippled or handicapped them for life. The tortuous route by which some of the 329 such children born in the British Isles with these defects received compensation by the Seagrams Distilleries is measure enough for the status of retributive justice for birth injury, even where the causal factor is known without a doubt. Similarly difficult legal avenues have been traversed by daughters of mothers who took diethylstilbestrol (DES) during pregnancy, and who subsequently developed a rare form of vaginal carcinoma. Although the hormonal damage due to DES has been known since the mid-1960's, it was only in September of 1979 that the first in-utero-damaged girl received compensation from the courts for her injuries.

In that month, the State Supreme Court in the Bronx, New York, awarded $500,000 to a 25-year-old woman who had undergone a hysterectomy and vaginectomy to treat a case of vaginal cancer linked to DES. The plaintiff won her case against Eli Lilly and Company, a pharmaceutical firm which was the largest manufacturer of DES during the 1940's and 1950's when this young woman was exposed. Although the plaintiff could not establish that she used a Lilly product, she nevertheless successfully sued Lilly on the premise that all manufacturers of a drug share responsibility for adequate testing and safety control procedures [1].

While these two cases are but the most sensational of a series of court rulings holding manufacturers and physicians responsible for informing their clients of risks of potential teratogenicity, they exemplify an increasing trend to recognize the vulnerable fetus as worthy of both protection and recompense for injury.[4]

JUSTICE AND UTILITARIANISM

A utilitarian approach to justice suggests that policies be shaped to the circumstances of nations and subpopulations as much as those of the developing fetus itself. Thus, while it is both prudent and just to regard the vulnerability of the developing fetus as deserving of special attention and care during the first trimester of pregnancy, such a perspective can deny a greater number of fetuses the minimal attentions needed to ensure a modicum of well-being and development. Often, the very medical advancements that are made possible by vast resources and wealth in developed countries place fetuses at risk of iatrogenic or other damage. Hence, justice for the good of the many may mean subordinating the claims of the few for scrupulous protection and, where desired, quality of their developing offspring to the prenatal needs of the poor who have few, if any, resources to provide even the most basic staples of pregnancy and prenatal care.

The rights of the fetus to protection are best seen then as conditioned rights, rights that are shaped by the larger social context in which mothers bear children. Where society assumes greater responsibility for the collective well-being of the population, as in those countries like Sweden and the United Kingdom which have socialized medicine, it appears appropriate to require equal treatment for all those who have entered pregnancy. But for countries like the United States where the distribution of medical care is still badly skewed toward the well-to-do, it appears appropriate to ask that justice mean differential treatment for those who are presently the least well-off. More significantly, retributive justice under those conditions would demand that injuries sustained as the result of past injustices (for instance, as the result of maternal deprivation or exposure to teratogenic substances) be recompensed by preferential treatment and, where appropriate, monetary compensation.

Thus, for prenatal life, justice is most meaningful when it regards fetuses as a class. Enlightened policies, in turn, are most likely to be those which address the least well-off group in society with regard to fetal outcomes and invests in their well-being accordingly. Such a conclusion means that prenatal diagnosis and other hand-tailored technologies should not be given priority until all fetuses at risk are afforded the opportunity for minimal well-being.

Department of Health Services
Berkeley, California

NOTES

[1] Leonardo da Vinci elevated the newborn as well as the fetus *in utero* to an object of great beauty and classic proportions.
[2] The precipitous cutback of Medicaid funds for abortion services in 1978 now precludes the universal availability of abortion.
[3] See [3] where D. Callahan describes this legal process in detail.
[4] Further rulings have since sharpened the definition of legal responsibility. In a March 20, 1980 decision, the California Supreme Court held that each manufacturer of DES was liable for the proportion of the judgment represented by its share of the drug market (Sindell v. Abbott Laboratories Sup. 163 Cal. Rptr. 132).

BIBLIOGRAPHY

1. Anon.: 1979, 'Victim Wins DES Case', *Modern Health Care*, September, 88.
2. Bannerman, R. M. *et al.*: 1977, 'Amniocentesis and Educational Attainment', *New England Journal of Medicine* **297**, 449.
3. Callahan, D.: 1975, *Abortion: Law, Choice and Morality*, Macmillan, New York.
4. Golbus, M. S. *et al.*: 1979, 'Prenatal Genetic Diagnosis in 3000 Amniocenteses', *New England Journal of Medicine* **300**, 157–163.
5. Jefferies, W. M.: 1979, 'Prednisone Therapy and Birth Weight-Correspondence', *Science* **206**, 96–97.
6. Kaufman, I. R.: 1979, 'Juvenile Justice: A Plea for Reform', *New York Times Magazine* (October 14), pp. 42 ff.
7. Reinisch, J. M. *et al.*: 'Effects of Prednisone Therapy on Birth Weight', *Science* **202**, 436–438.
8. Reinisch, J. M. *et al.*: 1979, 'Prednisone Therapy and Birth Weight-Correspondence', *Science* **206**, 97.
9. Smith, K. D. *et al.*: 1979, 'Prednisone Therapy and Birth Weight-Correspondence', *Science* **206**, 96.
10. Woollam, D. H. M.: 1964, 'The Effect of Environmental Factors on the Foetus', *Journal of the College of General Practitioners* **8**, Suppl. 2, 35–46.
11. Zamenhof, S. *et al.*: 1971 'DNA Cell Number in Neonatal Brain: Second Generation (F_2) Alteration in Maternal (F_0) Dietary Protein Restriction', *Science* **172**, 850–851.

ALBERT R. JONSEN

JUSTICE AND THE DEFECTIVE NEWBORN

Neonatology, the care of the critically ill newborn, is a new subspecialty of pediatrics. The first examination to certify specialists in neonatology was held in 1975. Even before this formal recognition of the specialty, ominous ethical clouds were gathering around it. Two years before, pediatricians Raymond Duff and A. G. M. Campbell reported that, in their newborn intensive care nursery, treatment was deliberately discontinued for some infants "with the knowledge that earlier death and relief from suffering would occur" ([3], p. 891). In the same year, a film, produced by the Kennedy Foundation, dramatized a case which had taken place ten years earlier: a newborn infant, suffering from Down's Syndrome, was allowed to die after its parents refused permission for life-saving surgery [8]. A quiet debate over the treatment of infants born with spina bifida had been going on during the previous decade and then broke into public view [6]. All this caught the attention of philosophers and theologians interested in the ethics of medicine, who then poured out a flood of literature on ethical issues in care of the defective newborn.

One question is constantly discussed in this literature: is it morally right to withhold or to discontinue treatment on the basis of the 'expected quality of life' of the newborn infant? Richard McCormick has answered affirmatively: "life is a value to be preserved only insofar as it contains some potential for human relationships. When in human judgment this potentiality is totally absent or would be, because of the condition of the individual, totally subordinated to the mere effort for survival, that life can be said to have achieved its potential" ([11], p. 175). Similarly, in the book, *Ethics of Newborn Intensive Care*, the authors propose that life-saving intervention could be discontinued if the infant receiving it "cannot survive infancy, or who will live in intractable pain, or who cannot participate even minimally in human experience" ([9], p. 144). In these circumstances, treatment may be withheld or terminated; the decision is justified by prospective quality of life of the infant.

Opponents of quality of life criteria sometimes criticize their use as a violation of justice. Warren Reich, for example, criticized McCormick's position: "the first objection against the 'rational' standard is that it is unjust, because it lacks equality, which is a fundamental characteristic of justice." The value of life, Reich contends, should not depend on whether the infant has "a specific quality — a certain level of relational potential — which is simply not shared by all and which is inherently variable in the way it is found in people, and that is not just" ([8], p. 503). Professor Paul Ramsey, among the most adamant opponents of quality of life criteria, scolds McCormick for his 'ever so slight' lapse into quality of life criteria. But he berates mercilessly the authors of *Ethics of Newborn Intensive Care* who had stated their dedication to "the conception of the equal and independent value of human beings." This conception, commonly cited as an expression of the basic principle of justice, the authors assumed as valid for their discussion and "judged that the burden of proof lies with those who would deny it" ([9], p. 145). They had in mind those authors who attributed no rights, or only diminished rights, to the newborn ([4, 10, 19]).

Professor Ramsey, with a thrust of his trusty irony, took up the challenge. "The burden of proof seems easily borne," he wrote; "first comes selectivity based on assessing the patient by his or her ability 'to respond affectively and cognitively to human attention or to engage in communication with others.'" Such criteria for selection belong to the category of "irrelevant or arbitrary reasons that deny the equal and independent value of particular patients" ([16], pp. 243, 256). The beleaguered authors, it seems, had erected a wall of principle which they themselves then breached. Selection on the basis of quality of life criteria, Ramsey contends, is fundamentally unjust.

Professor Ramsey does admit that there are limits to medical care for the defective newborn. When the dying process has already taken hold, therapy may rightfully and justly be terminated; medical efforts must turn from curing to caring. Other than the inevitability of death, only a 'medical indications policy' should dictate the relevant and non-arbitrary reasons required by justice. Such a policy would "direct attention to the objective condition of the (incompetent or voiceless) patient, not to abstract classification of treatments or to the wishes of any of the parties concerned" ([16], p. 159). Ramsey is concerned to transpose this medical indications policy "to the case of defective newborn infants without any theoretical mistakes from which flow grave injustice" ([16], p. 191). Presumably, those theoretical mistakes are made by ethicists who advocate, tolerate, or, in any other way, give credence to the invidious quality of life criteria.

Ramsey illustrates his medical indications policy by describing the practice of Dr. R. B. Zachary, an English surgeon specializing in the treatment of spina bifida. Dr. Zachary does not choose to operate on babies who, in his best judgment, will die within a few days; surgery will not halt the impending death. All other babies are operated upon unless an operation is surgically impossible, i.e., a very wide wound or very poor prospects for post surgical healing, etc. No judgment is made on the babies' future life and its qualities; if surgery can preserve some neurological function and reduce some handicap, it is medically indicated and, in Ramsey's judgment, ethically right.

The choice of spina bifida as the example of a medical indications policy is noteworthy. Ramsey offers one other example in passing, the correction of an intestinal obstruction in a Down's Syndrome baby. In this case, Ramsey points out, "if an operation to remove a bowel obstruction is indicated to save the life of a normal infant, it is also indicated treatment of a mongoloid infant" ([16], p. 192). These two cases have become the paradigm cases for the ethics of newborn care. Almost all of the ethical discussion has centered on the problems presented by these cases. However important this discussion has been, both for ethical theory and for practical decision, it misses certain features which are crucial to an understanding of the ethical problems of neonatology and of the problem of 'neonatal' justice in particular.

In the paradigm cases, the Down's Syndrome infant with duodenal atresia and the infant with spina bifida, surgery can be employed to save the infant's life, although a basic disorder would continue to afflict the life of the child so saved. In one case, the child suffers from a genetic defect which will retard mental development; in the other, the neurological damage done in the course of fetal development cannot be corrected by surgery after birth, leaving the child with major deficits, such as the inability to walk or to control the bladder and, perhaps, mental retardation. In both cases, the ethical question concerns the decision to perform surgery. If surgery was performed, the child would be likely to live, but with a life 'of diminished quality.' If surgery was denied, the child would be likely to die (certainly, in the case of the duodenal atresia), thus relieving it of a burdened life and its parents and society of it as a burden. The ethical question was posed in Shakespearean paraphrase: to treat or not to treat?

However, the majority of infants in need of neonatal intensive care are not afflicted with genetic or congenital anomalies. They are not usually candidates for surgery. They are infants born prematurely and of low birth weight (defined as less than 38 weeks gestation, and below 2500 grams, or 5½ pounds). Their primary problem is that their lungs are not yet mature enough to breath oxygen from the ambient air. This prematurity may result in severe deprivation of oxygen and subsequent acute and chronic lung disease. While prematurity causes a variety of problems, respiratory problems and its consequences are the most serious. At present, about 200,000 infants receive intensive care each year. Of these, about 75,000 suffer the problems of attending prematurity. Twenty-five thousand will die from respiratory problems, even with the best of care ([14], p. 249). This contrasts with the 5,000 born with spina bifida and the small proportion of the 3,000 Down's Syndrome babies who will require life-saving surgery ([1], pp. 73, 120). For the rest of this essay, I refer to the case of an infant so ill with respiratory problems as to occasion an 'ethical discussion' as the 'typical case.' The much debated, but less frequent ethical problem of whether or not to operate I shall call the 'paradigmatic case'.

The problem of justice appears differently in these two sorts of cases. Justice is taken to mean "treating similar cases similarly," or in a contemporary definition, "the equal treatment of all persons, except as inequality is required ... by relevant considerations or principles" ([5], p. 13). Authors who argue against quality of life criteria as unjust propose that they dictate dissimilar treatment for similar cases and do so because the 'consideration', viz., future quality of life, is not relevant to the decision at hand. In addition, they point out that these criteria, due to their indefiniteness, cannot be applied fairly: it is impossible, they say, to define these criteria in ways which will not be arbitrary ([7], p. 229). I am in general agreement with these criticisms when the case is a paradigmatic one; I do not believe, however, that they are appropriate criticisms against the use of quality of life criteria, in a carefully defined sense, in the typical cases. It is, I think, important to make this distinction for two reasons. First, the typical cases are by far the more numerous and deserve their own analysis. Second, practitioners dealing with the typical case should be aware that one form of quality of life criteria (which does, I think, reflect actual practice rather closely) is less problematic ethically than the form associated with the paradigmatic cases.

The first important difference, for the question of justice, between the typical and the paradigmatic cases, concerns equality. The authors of *Ethics of Newborn Intensive Care*, who were dealing exclusively with typical cases, affirmed 'the equal and independent value of human persons.' In the typical case, all infants born alive, with any prospect of survival, should be treated equally. This ethical principle is widely honored in modern pediatric and obstetric practice. The tiny premature who shows any signs of life and does not manifest any gross abnormality incompatible with life (such as anencephaly) is usually resuscitated. This standard of practice is, in some states, reinforced by law. Once in the nursery, all infants are considered equal: therapeutic efforts are provided in accord with the infant's individual medical condition.

Indeed, while equal consideration may be weakened by the emotions, exhaustion or prejudice of those responsible for care, a remarkable equality reigns in the nursery: an equality seldom found in later life.

In paradigmatic cases, discrimination threatens from the beginning. Some quality of the infant sets it apart from all other children and predisposes to prejudice. The presence of Down's Syndrome casts the shadow of prejudice on the infant. The presence of a lesion above the third lumbar vertebrae, with the attending prognosis for physical and mental development, places the infant in a category very likely to be 'prejudged' for non-treatment. If one can judge from Ramsey's polemic (with which, in these cases, I agree), one must marshall arguments against the prejudice and in favor of treatment. One must offer persuasive reasons why "if an operation to remove a bowel obstruction is indicated to save the life of a normal infant, it is also indicated treatment of a mongoloid" ([16], p. 192). The ethical arguments in the paradigmatic cases, at least in the current debates, seem directed toward dispelling the prejudice which immediately shadows the infant patient. In the typical cases, on the other hand, equal treatment is the standard of practice.

A second important difference arises from the temporal sequence of the decisions made in the typical and the paradigm cases. In the paradigms, the question is 'shall we operate'. It bears on a discrete action, surgical intervention. The medical indications for surgery and the probable consequences of surgery are relatively clear (clearer, of course, in the Down's Syndrome case than the spina bifida case). The answer must be yes or no; the decision and action must be taken in a circumscribed period of time. In the typical case, however, care of the infant is a trajectory along which many decisions are made and many events happen, related to each other in complex ways. Considerable energy, ingenuity, effort, time and money are expended on the care of the sick infant. Successes and failures, of major and minor import, follow each other. Indications for therapeutic intervention are not always clear; probably consequences are often very uncertain. How long do we wait?, how many times do we try?, what shall we do next?, are the questions most frequently asked in the typical case.

This temporal trajectory, during which a variety of interventions are made and the infant's condition waxes and wanes, is important for the ethical analysis of the typical case. First of all, it reveals a subtle, but critical difference in the way in which 'quality of life' is used in the ethical debate. In the paradigm case, quality of life refers to a set of behaviors and experiences which are almost certain to be the lot of the infant as it matures. Certainly, there will be degrees of difference: a diagnosis of Down's Syndrome cannot predict the degree of mental retardation, but can predict its inevitability. When quality of life is invoked in the paradigmatic cases, reference is made to these states of life; argument about the value of living in such a state constitutes a central piece of the ethical discussion.

In the typical cases, there is almost no similar certainty about the nature or the inevitability of the infant's future state. It can be suspected that certain evidence (and the evidence is often not so certain) of severe brain damage due to anoxia or intracerebral hemorrhage, foretells some retardation in mental development. However, the correlation is weak: some children with apparently severe damage develop normally; other children with little or no evident damage do suffer from retardation [20]. The clinical evidence shows, with greater or less certainty, some lesion. That lesion affects some function. If lungs are incapacitated by hyaline membrane disease, breathing will be difficult or impossible; if brain ventricles are filled with blood, cognitive and motor functions may not develop in a normal manner. Relatively little can be said about what this child's quality of life will be; much can be said about its present condition. Prognosis bears primarily on the possibility that the lung condition will clear, with or without treatment, or that the brain will not be too severely damaged by the hemorrhage. Prognosis about future development is speculative. Indeed, intuitive judgments, such as 'major hemorrhage equals major retardation', may be refuted in reality. Thus, reference to quality of life, in the typical cases, is much more speculative than in the paradigmatic. Indeed, quality of life references are somewhat like the 'fallacy of misplaced concreteness': what is concrete is the clinical evidence and projection of its immediate and long range course. As this projection moves away from the immediate, it becomes much less concrete. As it reaches the point of predicting ability to learn, it is extremely speculative. To backdate such speculation into the clinical situation is to pretend that the speculative is the concrete.

It is for this reason that such phrases as 'meaningful life' used, for example, by Duff and Campbell, should have a restricted usage in the typical case ([3], p. 76). It cannot be banished entirely, because some clinical evidence – total destruction of the cortex of the brain, for example – does correlate with a future life of severe retardation, limited to physical functions, sensory responses, and locomotion. If 'meaningful life' means a life with cognitive function, such an infant will not have a 'meaningful life'. The prospect for physical life which is self-sustaining, that is, with all essential organ systems sufficiently integrated that the infant can live independently of persistent supportive therapy, should be the primary consideration. Supportive therapy refers to all medical or surgical interventions which have as their purpose the restoration of natural function. If, after repeated efforts at providing supportive therapy, preponderance of evidence is against the prospect of self-sustaining life, supportive therapy may be discontinued.

The temporal trajectory of neonatal intensive care and its implications for 'quality of life' criteria show how decisions to terminate treatment meet the test of justice. At the beginning of the trajectory, all infants are given equal consideration. They are not treated identically, but each one is treated in relation to its medical need. Equality rests on the acceptance of each infant as worthy in and of itself; departures from identical treatment are justified by greater or lesser medical need. During the course of the trajectory, the perception of this need changes. Those responsible for its care watch the response to supportive therapy: some infants achieve adequate lung function without other problems; others achieve lung function but only after having sustained brain damage; still others fail to achieve lung function and continue to present new, difficult and sometimes intractable problems.

Need is one of the principles which, it is commonly allowed, justifies a departure from identical treatment of equal individuals. Philosophers argue over its role in theories of justice, but generally accept it as a valid principle. In particular, it has been proposed as the relevant principle in determining the just allocation of medical care [15]. The application of this principle to medical care is notoriously difficult. Clearly, a person who suffers cardiac arrest 'needs' resuscitation; does an aged person with a diagnosis of pancreatic cancer 'need' resuscitation during cardiac arrest? A woman who has lost a leg needs a prosthesis; does a woman who has lost a breast?, or a man who has lost his hair? These are not simple questions.

The premature infant who is brought to the intensive care nursery has a peculiar need. Its untimely emergence from the womb leaves it immature in many respects, of which lung immaturity is the most serious. Therapeutic response to the needs of immaturity consists in activities which will substitute temporarily for those functions which must still develop and, in some cases, foster their growth. Defined in terms of the clinical reality, the infant needs physical and physiological maturity sufficient to sustain its own life in the world. It needs functional integration of the vital processes. It 'needs' the medical means which will fulfill that 'need'.

Some philosophers have suggested that the principle of need fits into the theory of justice because it subserves the primary principle of justice, equality [17]. Actions in accord with the principle of need can be seen as attempts to restore those who suffer from some inequality due to natural or extraneous causes. Persons with such needs are at a disadvantage for no fault of their own. Obviously, response to need or to 'essential needs' does not *create* the equality envisioned by the primary principle. In view of that principle, persons are treated as equal, as the philosopher Frankena writes, "not because they are equal in any respect, but simply because they are human" ([5], p. 19). However, it would appear to be an implication of that primary principle that inequalities which put persons at a disadvantage due to no fault of their own should be redressed when possible. When provision of medical care is seen as a matter of justice, it can be interpreted as acting on the principle of need in order to redress the disadvantages arising from physical disability and to restore the rough equality which consists in persons having basic powers of action.

In neonatal intensive care, the rough equality at which therapy aims consists in basic integrated physical and physiological function with which the normal infant, born at term, enters the world. I propose that an infant who, after a reasonable course of therapeutic efforts, does not progress toward such a degree of integrated function, no longer *needs* intensive care. All care, except palliative, may be terminated. That termination is just: it no longer appears to be able to redress the need which called for its initiation, namely, the immaturity of essential vital function. The only remaining question concerns 'reasonable' and it is a question of prudence rather than of principle.

It might be possible to describe infants in this tragic situation as 'dying' and to say that therapy is no longer indicated for this reason. When this is obviously true in the clinical sense, the conscience of those who must 'turn off the respirator' is clear. In many other cases, it is less obviously true to call the infant 'dying'. The infant may, at some future time, 'pull through'. That future is just not easy for anyone to glimpse. Therapy has not yet failed; it is just not yet working. In addition, therapies may also be causing damage the longer or more intensively they go on. In these sorts of cases, Ramsey's preferred phrase, 'medical indications policy', seems uninformative. Certainly, gentamycin is 'indicated' for septicemia, but how long and under what conditions does it continue to be indicated? A respirator is ' indicated' when lungs cannot function; but for how long? Is it indicated when other organ systems, kidneys, liver, gut, begin to show evidence of disorder? Is it indicated when it may be implicated in causing tissue damage or brain hemorrhage? In these situations, 'medically indicated' is not a very illuminating answer to the question whether treatment should be provided. Any helpful

JUSTICE AND THE DEFECTIVE NEWBORN

answer would have to have, it would seem, some ethical presupposition hidden in it. That presupposition would bear on the desirability or value of the expected result of the treatment. Thus, a phrase like 'quality of life' seems needed as an explicit or implicit response to the question whether a treatment is medically indicated. No simple formula can sum up the content of 'quality of life'. Each infant presents its own problems. However, we have proposed that 'quality' be confined to the more immediate possibilities for integrated physical and physiological function rather than the far future of the child's growth and development. It does not seem to me invidious or unjust to discriminate between individuals who need or do not need therapy in this sense.

A further step must be taken to deal with one sort of medical problem which arises in the typical cases and presents the most difficult ethical problem. In this sort of case, therapy has succeeded in aiding the infant toward self-sustaining integration. Its lungs, heart, kidneys and other major organ systems — with one exception — begin to work in harmonious integration. The one exception is the brain. Since this organ can be seriously and sometimes irreparably damaged by deprivation of oxygen at birth or by major hemorrhage from poorly understood causes, the future 'quality of life', in the sense of mental development, is compromised.

Infants of this sort may be under treatment when this evidence becomes available. If intensive therapy for pulmonary or cardiac problems is still under way, there will be questions about continuing; if the infant has survived its major crises, but requires treatment of a curable, but life-threatening problem, such as infection or intermittent cessation of breathing, questions will arise about intervening. In one sense, the issue is the same as the paradigm case: should indicated therapy be withheld because of an 'irrelevant' consideration, namely, future quality of life? In another sense, it differs from the paradigms, since the implications of the clinically discernible damage can rarely be predicted with any accuracy. However, when damage is so great and so obvious as to lead experts to opine that prospects for the development of cognitive function are dim or non-existent, does that infant still 'need' medical therapy for other life threatening problems? If it were possible to define cognitive function and to say that potency for it was either present or not present, in an absolute sense and not in degree, I would say that such a child did not need life supporting therapies. If potency for cognitive functions will not be present or cannot be supplied, there is no medical obligation to sustain newborn life by medical therapies. Even if that potency could be predicted to develop only in a primitive way, leaving the infant with static sensory capacity, I would not argue for intervention in life threatening crises. In both cases, there is no possibility of bringing the infant to the basic equality which medical therapy can provide, namely, assuring the integral function of physical and physiological systems. The potency for cognitive life, rooted in cerebral structure, belongs to the basic equipment with which the normal newborn enters the world. Simply because the integration of those cerebral structures into psychic life is a future event does not, it seems to me, justify excluding it from consideration.

However, I am reluctant to use this argument in practice for three reasons. First, the state of the art permits accurate prognosis from cerebral damage to future development only in rare instance. Secondly, I doubt whether it is possible to define cognitive function or to identify criteria for the absolute presence or absence of its potency, in any way that would gain wide acceptance. Thirdly, given the impossibility of so doing, 'quality of life' judgments slip easily into the 'greater or lesser' which is the bane of fair and just application. Greater or lesser degree of cognitive function (predicted with greater or less accuracy) calls for a line to be drawn somewhere and it would appear that such a line will, at best, be arbitrary. Despite this reluctance, I have been convinced, in particular cases of gross damage, that the infant's future would fall outside the horizons with which, even on the broadest grounds, we might define the human. In such cases, I cannot consider it unjust to withhold life-saving intervention. In all other cases I consider that justice requires intervention.

This essay has dealt only with the ethics of justice as applied to the defective newborn. Medical care, of course, is governed by ethical principles other than justice. Arguments about duties, such as respect for sanctity of life, or about utility can also be applied to the issues of neonatology. It might be possible to demonstrate that the duty to respect life dictates a position much more conservative than the one I have developed within the context of justice arguments. It may also be possible to show that the principle of utility leads to a much more liberal conclusion. I have not attempted to report or to construct these other positions, nor to refute them (as I am inclined to do). Within the context of justice arguments, I maintain medical therapy is justly provided to newborns when that therapy, at its initiation, promises the establishment of a basic equality among all newborns. That basic equality consists in possessing those integrated functions of major physical and physiological systems (including cerebral capacity for cognitive life) necessary for sustaining life. It is not unjust to withhold therapy when it appears that this goal will not be achieved. This is evidenced by 'quality of life' considerations which are confined to the clinical situation and its immediate consequences; not by 'quality of life' of future development, except when that development falls outside the broadest definition of the human.

Readers who have reached the end of this essay may feel they have been cheated. They may or may not have been impressed by the arguments, but they might have found them irrelevant to the question that led them to read "Justice and the Defective Newborn." That question might have been one which many, on learning about neonatology, have asked: 'does this new technology expend great money and effort on saving only a few, while many others are left without medical care or other necessities of life?' The question may be one sometimes asked by those who have a more sophisticated understanding of neonatology: 'are we succeeding in saving many children from the ravages of prematurity at the price of keeping alive many others — who formerly would have died — with severe damage?' These questions inquire, in various ways, about the cost-effectiveness of neonatal intensive care. Cost-effectiveness in the provision of medical care arouses in many an 'ethical' anxiety. One critic recently castigated this approach as 'conveying a certain sense of the fast dwindling sanctity of life'.

This essay has addressed the cost-effectiveness question, but only obliquely. A direct approach would be premature. Much of the empirical data needed for a fair analysis is still unavailable and the interpretation of the data in hand is difficult. Several generalizations do seem justified. Neonatology does not expend vast sums of money only to snatch from death premature infants who will themselves be a burden on families and society. On the contrary, a yet unpublished analysis of all available statistics suggests that for the majority of prematures, especially for those weighing more than 1200 grams, the devastating effects of prematurity have been dramatically reduced [2]. In the past, most of these children would have lived and, during their lives, required special education or institutionalization. Today, most live without major deficits. For the much smaller population of tinier infants, the prospects are not so bright: they are saved from almost certain death, but often with physical and physiological damage which will burden them and others. However, to deprive them, as a class, of care would do little to promote cost-effectiveness, since they represent a very small proportion of the total population of prematures.

The approach taken here was oblique because it did not take up the question of whether it is just to benefit some while others are burdened. Rather it asked whether discrimination between the some and the others was justified by reference to 'quality'. The ethics of any cost-effectiveness policy would turn, it seems to me, on how one answered that question. Some have answered 'no' and, thereby, have placed stringent ethical limits on the range of any cost-effectiveness policy. Others answer 'yes'. They limit the range of cost-effectiveness by the manner in which they define 'quality'. The essay some readers might have wished to read under the title "Justice and the Defective Newborn" remains to be written. It can be written, in my opinion, only when two elements can be drawn together: a position on the fairness of quality criteria and the empirical data, actual or projected, about the costs and benefits of making decisions in accord with such a position. Neither of those elements yet exists with a clarity sufficient to allow their fusion into a theory of neonatal justice.

School of Medicine
University of California
San Francisco, CA

BIBLIOGRAPHY

1. Antenatal Diagnosis: 1979, Public Health Service, Washington, D. C.
2. Budetti, P.: 1979, Personal Communication.
3. Duff, R. S. and Campbell, A. G. M.: 1973, 'Moral and Ethical Dilemmas in Special Care Nursery', *New England Journal of Medicine* 292, 75–78.
4. Fletcher, J.: 1972, 'Indicators of Humanhood', *Hastings Center Report* 2 (November), 1–3.
5. Frankena, W. K.: 1962, 'The Concept of Social Justice', in R. B. Brandt (ed.), *Social Justice*, Prentice Hall, Inc., Englewood Cliffs.
6. Freeman, J. M.: 1973, 'To Treat or Not to Treat: Ethical Dilemmas of Treating the Infant with a Myelomeningocele', *Clinical Neurosurgery* 20, 134–146.
7. Grisez, G. and Boyle, J. M.: 1979, *Life and Death with Liberty and Justice*, University of Notre Dame Press, Notre Dame and London.
8. Gustafson, J. M.: 1973, 'Mongolism, Parental Desires, and the Right to Life', *Perspectives in Biology and Medicine* 16, 529–557.
9. Jonsen, A. R. and Garland, M. J. (ed.): 1976, *Ethics of Newborn Intensive Care*, University of California, Berkeley.
10. Lachs, J.: 1976, 'Humane Treatment and the Treatment of Humans', *New England Journal of Medicine* 294, 838–840.
11. McCormick, R. A.: 1974, 'To Save or Let Die', *Journal of the American Medical Association* 229, 172–176.
12. McCormick, R. A.: 1978, 'The Quality of Life, the Sanctity of Life', *The Hastings Center Report* 8, 30–37.
13. Miller, D.: 1976, *Social Justice*, Clarendon Press, Oxford.
14. Oh, W. and Stern, L.: 1977, 'Diseases of the Respiratory System', in R. E. Behrman (ed.), *Neonatal and Perinatal Medicine*, C. V. Mosby, St. Louis.

15. Outka, G.: 1974, 'Social Justice and Equal Access to Health Care', *Journal of Religious Ethics* **2**, 11–32.
16. Ramsey, P.: 1978, *Ethics at the Edges of Life*, Yale University Press, New Haven and London.
17. Raphael, D. D.: 1946, 'Equality and Equity', *Philosophy* **21**, 118–132.
18. Reich, W.: 1978, 'Quality of Life and Defective Newborn Children: an Ethical Analysis', in C. A. Swinyard (ed.), *Decision Making and the Defective Newborn*, Charles C. Thomas, Springfield.
19. Tooley, M.: 1972, 'Abortion and Infanticide', *Philosophy and Public Affairs* **2**, 37–65.
20. Volpe, J.: 1977, 'Neonatal Intracranial Hemorrhage', in J. Volpe (ed.), Symposium on Neonatal Neurology, *Clinics in Perinatology* **4**, 77–102.

MICHAEL D. BAYLES

JUSTICE AND THE DYING

Perhaps the primary question about justice and the dying is whether justice is the appropriate category for discussing most of the significant ethical issues concerning the dying. Many works concerned with the dying, even by non-utilitarians, emphasize considerations of benevolence rather than those of justice. The extent to which one believes justice is a major consideration depends on one's concept of justice. The following discussion is framed around three concepts of justice. The first section considers justice as respecting rights, focusing on the alleged right to die and a right to be told the truth. The second section discusses distributive justice and its possible conflicts with efficiency in the allocation of scarce resources between the dying and others. In particular, is the fact that people are dying a reason for not allocating resources to them? Compensatory justice, whether the dying are entitled to benefits to compensate for their dying, is analyzed in the last section.

I. FULFILLING RIGHTS

The relationship between justice and rights is more fully discussed elsewhere in this volume [3]. A general argument for equating justice and the fulfillment of rights is that fulfilling persons' rights respects them as ends. This claim has an obvious Kantian foundation, but Kant himself did not identify justice with respecting persons as ends. He held that respect for persons as ends grounds both duties of justice and duties of virtue ([5], p. 27). Kant restricted justice to the fulfillment of particular kinds of rights, namely, those which are capacities to bind others to external conduct ([5], pp. 34–35, 43).

Kantian rights of justice have two essential features. First, they impose correlative duties upon other persons. No self-regarding duties are duties of justice. Moreover, one must carefully distinguish claim-rights, which rights of justice are, from liberty-rights. A liberty-right makes no reference to other persons; it only indicates that conduct is morally permissible or that one has no duty to refrain from it. 'I have a (liberty-) right to do A' means only 'it is morally permissible for me to do A'. 'I have a (claim-) right to do A' means 'others have a duty not to interfere with my doing A'. Second, the duties of others correlative to a claim-right of justice pertain to conduct and not motivation or other subjective conditions. They are duties which may be enforced by law, whereas duties of virtue pertain to motives and, according to Kant, cannot be legally enforced.

Unless rights of justice are restricted in some such fashion, justice swallows the whole of morality. In the following discussion, Kant's first feature is the most important one. The second feature is rather peculiar to Kant's concepts of morality and moral worth.

The most discussed right of the dying is the alleged right to die. A right to die is usually conceived as a right to others' noninterference in one's dying. This concept of a right to die fits the Kantian concept of a right of justice, for it imposes a duty of conduct on others. The thrust of a right to die thus construed is against undesired efforts by medical personnel to prolong life. These efforts may be dramatic ones of the use of artificial respirators or the more simple administration of penicillin to combat pneumonia in a patient dying of cancer. The distinction between ordinary and extraordinary treatment (or other similar distinctions) is not relevant here.

Is this right a distinctive one of dying persons? Stated as a right to die, the answer appears to be yes. However, the underlying claim is to noninterference with one's body. As such, bodily integrity is the central value at stake. Medical personnel and people generally have a moral and legal duty not to interfere with or to touch one's body without consent. This principle underlies the laws of battery and informed consent and is not restricted to the dying. Consequently, the right to die as a right not to receive treatment without informed consent (or to withhold consent to life-prolonging treatment) is merely a specific application of the more general right to bodily integrity. It may still be important, but it is not a distinct and special right of the dying.

A right to die may also be asserted in a different context. Classical natural law theory (as well as Kant) held that one has a duty not to commit suicide, that suicide is wrong. Indeed, the distinction between ordinary and extraordinary measures was developed in this context. If a person had consumption and moving to the mountains would improve his health and prolong his life, would he have to do so or else violate the prohibition of suicide? The ordinary/extraordinary distinction was used to claim that moving to the mountains would be too much to expect, that failure to do so would not amount to suicide. One had a duty to refrain from actively taking one's life and to use ordinary measures to prolong it, but no duty to use extraordinary measures to prolong it.

A right to die could be asserted as a defense against the wrongness of suicide. As such, it is no longer a right of justice in the Kantian sense. Rather, the right to die merely asserts the liberty-right to, or permissibility of, suicide (or of not prolonging one's life). It imposes no corresponding duty on others; it simply denies that a dying person has a duty to prolong his life.

This point is more important than may at first appear. A natural response to the preceding argument is that if it is morally permissible for a dying person not to prolong his life, then others have a duty not to interfere with his dying. This response glosses over a significant problem. From the moral permissibility of X doing A (X's liberty-right to do A) it does not logically follow that others have a duty not to interfere with X's doing A (that X has a claim-right to do A). The first statement is solely about X's duties and states nothing about the duties of anyone else. However, that X's doing A is morally permissible is relevant to whether others have a duty not to interfere. If the dying had a duty to prolong their lives as much as possible, it would be much more difficult (but not necessarily impossible) to argue that they had a claim-right that others not interfere with their dying. Thus, the claim-right of justice to die receives considerable support from the liberty-right to die — the moral permissibility of not prolonging one's life or committing suicide.

A right to die can be taken in a third sense, as a claim upon others to take one's life in appropriate circumstances if one is unable to do so oneself. Consider a patient who is dying, in great pain controllable only by drugs which destroy mental alertness, and paralyzed from the neck down. Such a person may assert that he has a right to die which imposes a duty upon others to take his life. Such a right to die would be one of justice, for it would impose a duty of conduct upon others. However, the underlying rationale for such a right appears to be benevolence rather than justice. The point of such a right is to relieve persons of misery, a form of benevolence usually reserved for nonhuman animals such as horses and other pets. Few moral theories other than utilitarianism imply an affirmative duty to commit euthanasia. Instead of others having a duty to commit euthanasia, a more plausible principle is that their doing so is permissible in such circumstances. If suicide is permissible, then euthanasia is plausibly permissible when a person's condition is similar but he is unable to commit suicide. Taken as the permissibility of others committing euthanasia, a right to die is no longer a right of justice. It is not even a right of the dying person. At best, it is a liberty-right of others to end the life of the dying person, i.e., it is morally permissible for them to do so.

One other right of the dying deserves mention here, namely, the right not to be lied to. Much discussion of truth-telling has concerned telling cancer patients that they have cancer. The practice of two decades ago was not to inform patients [7], although the practice now is usually to inform them [8]. If truth-telling is based on a right not to be told lies (deliberate falsehoods or misleading half-truths), then it is a right of justice, for it imposes a duty upon others not to engage in certain conduct. Alternatively, truth-telling may be construed as a claim-right to be told the whole truth, a claim to full disclosure. This right is also one of justice, for it imposes an affirmative duty upon medical personnel to disclose fully information about a patient's condition.

By this point, the difficulty in taking justice to the dying as fulfilling their rights should be apparent. It encompasses a large number of duties to the dying, such as truth-telling, which are not usually thought of as duties of justice. As Aristotle noted with respect to justice as conformity to law, justice as fulfillment of rights swallows too much of morality. One needs a more particular concept of justice.

Following Aristotle, the rest of this essay is concerned with justice as the more particular concept of treating equals equally and unequals unequally in proportion to their relevant differences ([1], p. 1131a–b). Justice of this sort does found claim-rights, but not all claim-rights are rights of justice.

II. DISTRIBUTIVE JUSTICE

Perhaps the central issue in the more particular Aristotelian concept of justice is whether the fact that one is dying is a relevant difference. In the context of the distribution of care and resources, that a person is dying is often argued not to be a relevant difference. That is, that a person is dying is not a reason for excluding him from receiving certain benefits. This claim is considered in this section under the heading of distributive justice. Alternatively, that a person is dying may be claimed to be a relevant reason for providing him certain extra benefits. Although this too is an element of distributive justice in as much as it concerns the distribution of benefits, it is discussed in the next section under the heading of compensatory justice. Claims to extra benefits for the dying may be seen as compensation for their condition as opposed to compensation for wrongs done to them. Nothing depends on this terminology; it is used merely to separate the two issues of whether dying is a reason for or against receiving benefits.

Assume that health care should be distributed equally amongst all persons unless there is a relevant difference between them, medical need being the most obvious such difference. One question is whether that a person is dying is a good reason to provide him a less than equal share. Consider nursing care

JUSTICE AND THE DYING 113

as a resource on a hospital wing. Should nursing care be distributed equally amongst all patients, those who are dying and those who are not, simply on the basis of medical need? A nurse remarked to me that nurses would devote more time and attention to patients who were not dying. Nurses, like other health care professionals, this nurse maintained, are trained to cure and to produce health. While basic care would be provided the dying, nurses would be inclined to devote extra care to patients with a prognosis of recovery. Other nurses doubt the accuracy of this factual claim. Whether or not it correctly describes nursing practice, it raises the issue of whether such a practice would be unjust.

Several lines of argument are possible. First, one may claim that nurses have no duty to provide more than basic care. No patient has a complaint should a nurse not provide care above the basic. To the extent extra care is provided, a nurse has the moral liberty to bestow it as a gift upon whomever the nurse wishes. Choosing not to provide extra care to the dying would not be unjust.

Second, one may argue that nurses are hired to provide care and that all care provided while they are on duty is part of the job. Any extra care which they are entitled to bestow as they wish must be given off duty. Care which is not medically needed to restore persons to health, for example, conversing with patients, washing and shaving them, and so on, is directed to the patients' present needs and well-being. Dying persons have the same present needs as nondying persons, so they are equally entitled to such care.

Third, one may argue that even such care gives patients a feeling that someone is concerned for them and that their lives are worth living. Hence, it may contribute to patients' desires to live and thus increase the chances of survival and speed of recovery of those who are not expected to die. In effect, the extra care provided nondying patients is medically beneficial. There is no point giving penicillin to a patient when it will do no good or extra care to speed recovery when recovery is not likely. Consequently, providing extra care to the nondying is not only permissible but required by the principle of allocating care on the basis of medical need. Conversely, that a person is dying indicates that he does not medically need the extra care and so is a reason for not providing certain benefits.

At this point one must be careful not to broaden the concept of justice so as to incorporate all moral considerations which may be relevant to a particular allocation of benefits. If one does so, then no morally justifiable allocation can be unjust. Justice is then equivalent to morally right and is not simply one of several moral principles or considerations. If justice is one of several moral considerations, then it may be outweighed by the others so that an act may be unjust but morally right. In particular, justice may conflict with utility (efficiency, productivity).

The problem is that the concept of medical need is vague enough to encompass all relevant moral considerations. A clear case of lack of medical need would be penicillin for a person with a cold. As it has no effect on the cold virus, it will not do any good. However, in other cases the concept of medical need is unclear. Do dying patients need to be resuscitated should they suffer cardiac or respiratory failure? Some patients are 'no-coded', that is, resuscitation will not be attempted should they arrest. One might say resuscitation is not medically needed because even if revived, they will not survive long. Nonetheless, resuscitation would accomplish its immediate aim, namely, revival. The concept of medical need is here incorporating a kind of utility, the utility of longer life. The concept of medical need is not a purely biological one but includes value considerations.

Generally, the dying will benefit less than the nondying from certain types of medical treatment simply because they will not live as long. The concept of dying is a bit vague. Omitting cases such as miners trapped by a cave-in, three conditions are required for a person to be dying ([4], p. 35). (1) The person has an irreversible illness or exhibits a deterioration known to lead to death. (2) Good reasons exist to believe the person will die of that illness or deterioration. (3) Death is likely to occur soon. While this third condition is vague, in conjunction with the other two, it helps limit the range of persons who are dying. People are not dying simply because they are progressing toward the termination of their lives, yet the dying include more than those with but a few minutes to live. A person with gastro-intestinal cancer is dying if the disease is progressing even though the person may be expected to live another year. A person with an organ transplant but exhibiting no signs of rejection is not dying even if the five year survival rate should be less than 50%.

To the extent the concept of medical need incorporates benefit over time, then that a person is dying is a relevant reason of justice not to provide certain types of health care. By definition, the dying do not have much expectable life and cannot receive long-range benefits. If one does not incorporate a consideration of benefit into the concept of medical need, then not providing the health care is unjust. However, it may still be morally justifiable, because considerations of utility and benevolence may override those of justice.

Another issue of distributive justice and the dying which cannot be considered in this paper is how one chooses among the dying when insufficient resources exist to provide care for all. This issue concerns such matters as selecting patients for dialysis when insufficient machines exist to treat all of them and is discussed by Rescher [9] and Childress [2].

III. COMPENSATORY JUSTICE

Is the fact that a person is dying ever a reason for providing him extra health care? Commonsense seems to support such a claim. Friends and relatives often feel that because a person is dying, they have a special obligation or duty to visit the person and do things for him. This duty is greater than if the person were merely ill and expected to recover.

Commonsense feelings are not an accurate guide to moral thinking. They require analysis and ethical appraisal. One reason people often feel such a duty or obligation toward the dying is that they will not have much more time to make amends for prior neglect. They do not want to compensate the person because he is dying but rather to atone for wrongs (at least of inattention) committed before the person was dying. While the person's dying is perhaps the precipitating cause of people devoting extra care, it is not the reason or justification for it. Another reason for extra care is that little time remains in which to express one's love or friendship to the person. The brevity of time is due to the person's dying, yet a duty to express one's love or friendship for a person (not *to* love or be friends with a person) pertains irrespective of whether that person is dying. These feelings should be expressed as the other needs reinforcement or assurance of one's affection. (Of course, they may also be expressed when no such reassurance is needed and such spontaneous expression may be more indicative of moral worth than their expression from a sense of duty. Still, one may have a duty to express these affections when one is not presently inclined to do so.) The dying, because they know they are dying, may have a need for, and therefore a claim to, expressions of affection. Nonetheless, these expressions are not ordinarily considered part of health care, at least that part provided by medical personnel. Moreover, as the dying cannot claim (as a matter of justice) that medical personnel love them or be their friends, these points do not apply to them.

The reverse of the argument in the previous section for nurses having a duty to provide extra care and attention to the nondying might be used here. The dying are more insecure and psychologically disturbed than the nondying. Because medical care should minister to the whole patient and not just biological mechanisms, the dying have a claim to more personal attention than the nondying in order to improve their mental state. "And if greater needs call for greater care and concern, then the dying deserve more, not less of it, than the healthy" ([6], p. 58). This argument assumes that it is equality of benefit or outcome (not resources *per se*) which is relevant. It also rests upon the (probably correct) statistical claim that the dying need more psychological reassurance than the nondying, a claim that can be incorrect in particular cases. Thus, the connection to the condition of dying is empirical and not conceptual.

Do the dying have a special claim to nonpsychological resources such as pain medication? The answer is no. Pain medication is provided to relieve pain, which is the relevant indication rather than dying. Nor does a strong empirical correlation appear to exist between dying and having pain. Many persons have and do die rather peaceful and painless deaths.

IV. CONCLUSION

If justice is the fulfillment of rights which constitute a claim upon the conduct of others, then it comprises most of what medical personnel and others owe to the dying. However, many of these correlative duties are unrelated to a person's dying or justice as ordinarily construed. They may be grounded in benevolence, bodily integrity, and other considerations rather than justice. Moreover, the belief that the whole or most of the relevant moral considerations for care of the dying is captured by the concept of justice may rest upon confusions between claim-rights and liberty-rights.

A more particular concept of justice is to treat equals equally and unequals in proportion to relevant differences. If justice requires that health care be distributed solely on the basis of medical need and medical need includes benefit over time, then that a person is dying may be a reason of justice *not* to provide health care that ministers to the future well-being of the patient. The strength of the reason is proportional to the person's expectable remaining life. Alternatively, considerations of utility may here outweigh those of justice. In either case, the connection between dying and health care is conceptual, because the concept of dying entails that a person will be dead soon. That a person is dying may also be a reason for providing extra health care to the extent it ministers to the patient's present psychological and physical well-being. However, for claims of this sort to be shown, one must establish strong empirical correlations between dying and the condition to which the care ministers. Most correlations of this sort will probably pertain to psychological rather than physical well-being.

Justice, however, may not be the most important moral category for considering health care for the dying. All conceptions of justice focus upon the rights of the dying.

Rights are generally appealed to more readily (and the adversary relationship is more typical) with strangers than with those we are close to such as family or friends. But certain situations or states tend by nature to be very personal, very private, and the kind of relationships particularly needed at that point are close, personal and understanding ones. One such state is that of dying, and one group with whom one needs that kind of relationship is the medical staff. For anyone, patient or staff member, to determine whether and how to treat more or less exclusively on the basis of rights would be to risk turning this very private and personal experience and relationship into an adversary and public one ([6], p. 143).

Westminster Institute for Ethics and
Human Values
Westminster College
London, Canada

BIBLIOGRAPHY

1. Aristotle: 1915, *Ethica Nicomachea* (transl. by W. D. Ross), in W. D. Ross (ed.), *The Works of Aristotle* 9, Oxford University Press, Oxford.
2. Childress, J. F.: 1976, 'Who Shall Live When Not All Can Live?' in R. M. Veatch and R. Branson (eds.), *Ethics and Health Policy*, Ballinger Publishing Co., Cambridge, Mass., pp. 199–212.
3. Golding, M. P.: 1980, 'Justice and Rights: A Study in Relationship', in this volume.
4. Gorovitz, S.: 1978, 'Dealing With Dying', in M. D. Bayles and D. M. High (eds.), *Medical Treatment of the Dying: Moral Issues*, G. K. Hall & Co., and Schenkman Publishing Company, Boston, pp. 29–46.
5. Kant, I.: 1965, *The Metaphysical Elements of Justice: Part I of the Metaphysic of Morals* (transl. by J. Ladd), Bobbs-Merrill Co., Indianapolis.
6. Keyserlingk, E. W.: 1979, *Sanctity of Life or Quality of Life*, Study Paper, Protection of Life Series, Law Reform Commission of Canada, Ottawa.
7. Oken, D.: 1961, 'What To Tell Cancer Patients: A Study of Medical Attitudes', *Journal of the American Medical Association* 175, 1120–1128.
8. Novack, D. H. et al.: 1979, 'Changes in Physicians' Attitudes Toward Telling the Cancer Patient', *Journal of the American Medical Association* 241, 897–900.
9. Rescher, N.: 1969, 'The Allocation of Exotic Medical Lifesaving Therapy', *Ethics* 79, 173–186.

SECTION III

ISSUES OF MACRO-ALLOCATION

H. TRISTRAM ENGELHARDT, JR.

HEALTH CARE ALLOCATIONS: RESPONSES TO THE
UNJUST, THE UNFORTUNATE, AND THE UNDESIRABLE

I. INTRODUCTION

In this essay I examine different ways in which one can view health care allocations while trying more to determine what the relevant questions are than to propose particular answers. As a consequence, (1) I do not endorse a particular system of health care allocation nor (2) do I advocate a single ethical theory that will allow one then to derive a system of health care allocations. Rather, I address some of the bothersome difficulties that attend most of the ways one can view health care distributions with respect to ethical obligations. Towards this end I will contrast three genre of systems. There are surely many gradations, but I have chosen these three, for I believe they cover the spectrum and sufficiently signal the interesting issues.

1. The Pure Free Market System
In a pure free market system, one gets what one can pay for and what someone is willing to sell. The free market system is based on the trading of some goods for health care goods. It turns on the notion that one can exchange the resources one owns for resources that other people own — a notion that works strongly against egalitarian schemes.[1] This has been the prevailing system in North America.

2. Mixed Systems
A partially free market for health care with a decent minimum of health care for all.

In a mixed system (as I mean it here) a decent minimum is provided for all independently of their ability to participate in the market, and in addition, one can purchase additional care if one has the resources, and if someone is willing to sell the services. The second system, which is a two-tiered system, I will refer to as a health insurance scheme, where all are insured against a certain level of costs and are *assured* that they will receive a certain minimum amount of treatment. It is obvious that such schemes can range from a Medicaid-Medicare system as we have today ([8], [10], pp. 101–104, 133–135, [11]), to a totally comprehensive health insurance scheme, such as a number of proposals that have been introduced in Congress.[2]

3. The Egalitarian System

In a strictly egalitarian system everyone would receive the same standard of care, presumably one that affords at least a decent minimum, and there is no possibility to acquire additional health care on the open market. The third option underlies the notion of an inclusive health service, which attempts to encompass all health care providers and health care recipients. A second system, a second tier, is, under such a view, held to be improper and is not tolerated. Thus there have been arguments in Britain against allowing a second free market tier ([1, 2, 4]).

One must realize that one approaches the analysis of the issues in the allocation of health care through a sea of rhetoric raised by such competing views. Protagonists of various views strongly claim rights and assert what is just or fair:

I have a right to health care.

I have a right to health care, and what I mean by that is the provision of such services as appendectomies, treatment of myocardial infarctions, etc.

I have a right to health care, and that should include free rhinoplasties, mammoplasties, abortions on request, etc.

It is unfair that the rich can have better rhinoplasties than the poor.

It is unfair that the rich should have rhinoplasties for frivolous reasons while the poor can have them only for stricter medical criteria.

It is not fair that he should have leukemia.

It is not fair that she should have leukemia and not be afforded adequate treatment.

It is not fair that he should have leukemia and not be afforded the very best treatment.

What is at stake in such claims is the meaning of terms such as 'rights,' 'justice', 'fairness'. For example, when someone says that "X is morally wrong," it usually means that he or she disapproves of the circumstances. The philosophical issue is the moral force of the disapproval invoked in condemning a particular form of health care allocation. In short, what is at stake is the moral cash value of ethical claims regarding different schemes of health care allocation. For example, when I assert I have a right not to be shot without my consent, what is the force of the term 'right'? Does it mean that shooting me under such circumstances is not compatible with the attacker being a member of the moral community? Does it mean that shooting me will act to decrease the balance of value over disvalue? Does it mean that shooting me is incompatible with some ideal of beneficence or non-maleficence? Or does it simply mean that I find the prospect unappealing and would like to persuade you not to do it? Similarly, how strong is the claim to a right to health care? What will count as proper allocations of health services, and of goods for health services, will turn on an analysis of such issues. If rights to health care are not equivalent to interests in particular goods and values, but depend on the very notion of a moral community, then allocation under them will in part exist independently of goals.

The determination of the significance of ethical claims I take to be one of the important services of ethics. That is, ethics functions properly as an attempt to reconstruct as best one can the presuppositions ingredient in various kinds of moral claims. As such, ethics is a rational enterprise in which one defends or criticizes ordinary moral claims and sentiments in terms of their ability to hold together coherently or in terms of their incoherence both among themselves and with other presuppositions about the world and our actions. One must also tread carefully so that one gives a rational reconstruction and does not simply rationalize one's own pet inclinations. Alasdair MacIntyre, for example, has remarked upon the sad fact that the Hawaiians did not have ordinary language philosophy at their disposal in order to give an account of their taboos and thus perhaps preserve them [9]. According to Captain Cook, his men were puzzled by the fact that ordinary Polynesians knew that it was offensive to eat with members of the opposite sex, though it was enjoyable to sleep with them, so that Polynesians of sensitive moral conscience would have felt shame, indeed guilt, for having eaten with a member of the other sex, but not for having slept around somewhat promiscuously by Puritan English standards.[3] In general, our moral intuitions are different from those Polynesians. I take it what we want to do here is to determine which moral intuitions make sense with regard to the distribution of health care resources without simply ratifying our ordinary inclinations.

Finally, insofar as we are seeking an answer to a rational question we have eschewed the use of force. That is, it is one thing to ask how to force on others the pursuit of the goods one embraces, and another to decide to determine what reasons justify the pursuit of which goods. Insofar as one wishes to view the moral community as one based on mutual respect and persuasion only through reason-giving and peaceable manipulations, not force, one must eschew the use of force except in self-defense or in other circumstances in which the person subject to the force has explicitly or implicitly agreed to the duress. As such, respect of freedom becomes a necessary condition for the possibility of an ethical community, a side constraint, as Robert Nozick has put it ([12], pp. 30–34). By side constraint I mean that one has to respect freedom first, and after that one can, for example, be an utilitarian without violating the basic notion of a moral community in the sense of a community not based on force. Ethics as an applied enterprise thus becomes the logic of a pluralism, a means for negotiating moral intuitions by the use of reasons, not force, within the bounds of mutual respect of freedom. Insofar as one eschews force as the basis of the community, one must tolerate the vicious choices of others insofar as those choices do not involve the use of unconsented force against the innocent.

II. BASIC PUZZLES

Reflections on moral claims to health care allocations reveal a number of fundamental unclarities. The first concerns how to rank claims to health care versus other claims; the second concerns how to rank various kinds of claims to health care. These must be examined prior to judging any concrete claims of rights to health care.

1. Allocations of Goods Among Health Care and Other Enterprises

Before one considers the allocation of health care resources among those wanting health care resources, one must first inquire why and to what extent health care should be given a priority over other goods. Obviously, the health care enterprise competes with other enterprises, for example, art museums. One may argue that rights to access to art museums, formal gardens, good wine, and good philosophy should take precedence over rights to health care. Better to have a more pleasant life of shorter duration, than a longer one of less quality. In part this is Plato's point in the Third Book of the *Republic* (3.405–408) where he argues for the provision of preventive medicine (gymnastics) and acute health care when it promises to return one to a functional state, but not for chronic health care. Not only would it be too costly, but it would not afford individuals the reward of a life contributing to the polis.

2. Needs vs. Desires

Even if one decides to pursue clinical medicine, one will have to decide which programs are more pressing. Since human life is finite, one could invest an indefinite amount of energy and resources in blunting the vexations to which we are subject, a point illustrated in the modern television myths of the six-million-dollar man, woman, and dog. The possible allocations to health care

are potentially indefinite in that one could always wish to live longer, though there are limits to the amount of possible allocations to food, in that there is only so much one can eat. One needs to rank being healthy against other values. And there are different senses of achieving health that must be ranked. How, for example, does one rank the value of preventing different kinds of death. Is it one's concern to prevent early death, natural death, late death? When is one living long enough so that one's claim to be kept from death is weakened? Or how does one rank preventing death vis-à-vis the value of preventing pain or disability? How does one rank the importance of curing leukemia with the importance of blunting the pains of arthritis? And how does one compare the pleasing things that medicine does: rhinoplasties, mammoplasties, and non-therapeutic abortions, with those aimed at reducing mortality and hard core morbidity? Those making allocations of health care resources will have to decide (at least implicitly) which needs are more pressing and which are more frivolous.

3. The Unfair vs. the Unfortunate

But before one can address what would be a reasonable priority in the allocation of resources, one must ask when *must* one provide resources for health care? Here I have in mind trying to understand when it is fair and when it is unfair to fail to provide health care. I take it we have some rather clear cases of unfairness. If someone becomes ill because of negligently prepared food, there is a sense in which those who are responsible for the illness can be said to be unfair or unjust if they do not give or pay for treatment. What, though, if someone is walking along and is hit by a falling tree, which no one could have reasonably known was rotten, and one is an uninvolved bystander and declines to give aid? Is that unfair? Is it an unjust society that makes no provision to help such persons, but expends its funds instead in support of art museums? Who is responsible for the tree falling? We say it is an act of God. So I suppose God is responsible, if anyone. What if an individual (or the society) is standing there and does not help? Is it unfair or unjust? I take it that that depends on whether one thinks there is a fiduciary relationship between the person injured by the act of God and the bystander, or the society in this case.

We might very well say that the individual who did not render help, if he or she could have done so without great inconvenience, or the society which does not develop a system for such aid, if it could have been provided without great inconvenience, is shameful, and not the kind of community in which one would like to live. But is it unjust? I take it that in part turns on whom one thinks owns what and in what way (a point I will address below). My point here, though, is to suggest various attitudes one might have to the natural lottery: the workings of chance that give some long life, others painful and early deaths, yet others years of compromised existence, etc. Health care allocation systems exist in part as responses to the natural lottery. Consider the moral attitudes one could have to the natural lottery.

(a) The natural lottery is morally neutral.

One might hold that the products of the natural lottery are marvelous, indifferent or terrible, but no one is responsible, that is, unless one thinks God is responsible. Some people turn out to be handsome or beautiful, some to be plain, and others turn out to be homely. Some turn out to be smart, and some do not. Some are born healthy and others are born with incredibly painful diseases. The beautiful are desired, the intelligent have advancement, and the ill and disabled are hindered and suffer. However, under this construct, there is nothing unfair about it. It is awful, ugly, and evocative of sympathy, but it is not unjust. Thus, we may not be feeling or decent people if we do not respond, but we are not unfair or unjust. A society may be a despicable one if it does not respond with aid to those in dire need, but it is not an unjust society in denying something other individuals own to those in need who do not agree to provide for those in distress. In this regard one should remember a usage of justice in the Bible: "Do not pervert justice by giving false awards, whether by taking a man's poverty into account, or by flattering the great; give everyman his just due." (*Leviticus* 19:15)

(b) The differences in the results of the natural lottery are morally significant, but in a reverse fashion.

Such views see the moral life as an attempt to render human life rational and to blunt unfortunate outcomes by treating them as circumstances of unfairness or injustice. These views usually turn upon an ideal observer or group of ideal observers. One might think here of John Rawls's participants in the original position, who like demigods decide what would be an acceptable distribution of social goods. In such circumstances it is reasonable to ask, were we to view reality from that viewpoint, whether it would be fair or just to abide by the natural lottery in distribution of social goods. And the answer is no, given Rawls's thin theory of the good [13]. I take it that the difference between John Rawls and Robert Nozick is the extent to which we should see ourselves in the image and likeness of God and assume that special perspective. Depending on how one answers that question, one will see different moral obligations to distribute health care.

The sense of the duty to allocate resources to health care is different depending on which viewpoint one takes, for if the natural lottery is morally neutral, then providing inadequate health care is not *prima facie* unfair or unjust, though it may be indecent and unfeeling. However, if we view the world and the natural lottery's distribution of social goods as it should have been, were it to have been structured in order to support the moral order, such differences should, in justice and fairness, be obliterated as far as is reasonably possible. Where one draws the line between what is *unfair* and *unfortunate* will, as a result, have great consequences as to what allocations of health care resources are just or unfair as opposed to desirable or undesirable. If the natural lottery is neutral, in the sense of not creating an obligation to blunt its effects, one does not have *prima facie* grounds for arguing for a right to health care on the basis of claims of fairness or justice. If one attempts through the fabric of morality to set aside the effects of the natural lottery, one is claiming a right to health care as an issue of justice or fairness on the basis of what is tantamount to a right to health. Being denied health under such a view is a circumstance that remains unfair or unjust unless society acts to remedy it, which would provide a strong ground for an encompassing system of health care.

4. Owning Goods

One, however, cannot make sense of the implications of the natural lottery without also coming to terms with some account of the entitlement to goods. There are many views of what it means to own things, and I can here at best be only sketchy. What I wish to do is to contrast a position in which one is said simply to own things, with one in which one has a right to the goods one possesses, given a justification of the distribution of those goods by appeal to an ideal distribution, as occurs in John Rawls's *Theory of Justice*. Notice, if one owns things, the onus is on those who would want to take them away. However, if possession turns on a notion of an ideal moral distribution and that ideal is one of equality of distribution, or the most equal distribution compatible with the greatest amount of goods for the least-well-off class, then one may increasingly be pressed to justify why one happens in fact to have more than others. The first, i.e., ownership as an extension of the freedom of the owners, justifies possessions historically in terms of the propriety of their acquisition and transfer; the second, i.e., ownership based on accordance of the actual distribution of goods with an ideal distribution of goods, justifies them by appeal to a desirable goal of distribution.

To return to the example of the person injured by the tree: if one owns one's resources, there is an important sense in which the person in need of care has no right to those resources. If one is not forthcoming with assistance for the needy, others may have good grounds for saying that one is hardhearted or unfeeling, but not that one has been unfair or unjust. In such circumstances, the misanthrope, or a society that made such a policy the basis of its health care, would be unfair or unjust only if the misanthrope tripped the injured person, or the society otherwise prevented the injured from having access to health care. That is, one might have negative rights to health care, but not positive ones, in the sense that there is a difference between a negative right to life, the right not to be shot without one's consent, and positive rights to life, the right to sufficient goods to sustain one's life. If there are only negative rights to health care, then the right to health care becomes a right to form a contract for health services. The right is, however, simply a negative one; a right not to be interfered with in forming contracts.

However, if we do not in a basic sense own goods, but rather are entitled to a certain distribution of them because that scheme of distribution produces, let us say, the greatest balance of value over disvalue, or leads to the benefit of the least-well-off class, then one may indeed have acted unfairly or unjustly, or the society may have been unfair or unjust in not being forthcoming with resources, in order to establish a proper pattern of distribution of goods. The right would, on such grounds, be a positive right in virtue of the desired pattern of distribution.

III. SYSTEMS OF HEALTH CARE DISTRIBUTION

These considerations have import and implications regarding how one views the morality of various systems of allocating health care resources.

1. The Free Market System
The free market system comes very close to the contractual scheme of health care resource allocation that has been the paradigm for most of North America, and has been built on a right to that health care that one has the money to buy, and that someone is willing to sell. Moreover, in order to avoid being unfeeling, physicians have given away some services free, and the wealthy have endowed eleemosynary hospitals to provide care for those who have lost out in the natural lottery. If
    (a) the natural lottery is morally neutral, and
    (b) people own their goods, and
    (c) society is simply a grand negotiated contract, then a free market system

is morally licit — and, if the system lives up to its own ideals of charity, it may be fairly generally acceptable.

2. *Mixed Systems — Two-tiered Systems*
Two-tiered systems, such as those proposed by Charles Fried, are obviously open to various modes of justification [6]. Let me suggest only a few at this juncture. One might be John Rawls's theory as modified by Ronald Green in which health care is listed as one of the primary social goods.[4] Green suggested how one could reconstruct via Rawls a sense of justice or fairness in the allocation of resources so that a pattern of health care allocation is justified by an appeal to a pattern of allocating goods. Green and Rawls are asking what the grammar of talk about justice or fairness is like — they are attempting to outline the necessary conditions for the possibility of a grand practice, or a collage of practices — being fair in distributing resources. In short, they are giving a transcendental account, an account of what views of goods and entitlements are required in order to make a certain sense of justice possible. It is important to understand that theirs (i.e., Rawls's and Green's) is a sense of justice that turns in great part on how to avoid being exploited. It starts with the question of how people would divide resources so that they would all agree to the reasonableness of taking whatever amount came their way. It is like asking, how would one distribute health care resources so that one would not mind getting any particular allocation. Notice that such a view presupposes that there are no particular entitlements, rather than one can justly divide resources without asking the permission of the property owners. Obviously, if someone owns things, as Robert Nozick holds that one owns things, one cannot distribute privately owned resources without the permission of the owners. In contrast, for Rawls and Green fairness is allocation by blindfold, that is, the blindfold of the original position (i.e., that one does not know one's place in society, one's natural assets and abilities, one's conception of the good, the special features of one's psychology, the particular circumstances of one's society, or the generation to which one belongs) ([13], p. 137).

In the case of health care distribution this would mean at least that one would want that distribution of health care resources that would not disadvantage one, should one turn out to be a loser in the natural lottery of health and wish a lot of health care. However, since one would not be envious (Rawls's and Green's blindfolded demigods are not envious) ([13], p. 143) if someone got better health care, as long as the status of the least-well-off was maximized, one would tolerate a two-tiered health care system. One would in general be better off absolutely, though perhaps not relative to others, no matter what happened in the natural lottery, given a system designed to maximize the health care position of the least-well-off class.

The difficulty is, however, how to characterize the least-well-off class. If it is in terms of the least-well-off with regard to primary social goods generally — then, such a distribution is perhaps more manageable. Health care allocations would be understood within a more general allocation. But if it is to mean the least-well-off class with respect to the natural lottery of health, the puzzle may be irresolveable. Some may be so badly off that little one does may help their health status. For them, what distribution of health care would be a just distribution? Unlike feeding the hungry, treating the ill might not raise the ill in all cases to an acceptable level of health. If that is the case, what is the standard for a just allocation of health care? Could it be an equal distribution when some need so much more, and others so much less? Also, if distributions of holdings are just only if they are to the advantage of the least-well-off class, what of those who have two kidneys when others have none? Do they have a right to the extra kidneys if entitlements are in terms of just, ideal distributions?[5] It is puzzles such as this which make a two-tiered system an inviting solution. Namely, one should provide all with a decent minimum amount (a created criterion), and otherwise allow individuals to contract for further services.

3. Egalitarian Systems

How, though, is one to understand egalitarian health care systems, as for example forwarded by individuals such as Robert Veatch [18]. It is here that many of the strong implications of the slogan, *equal access to health care*, become apparent. In the case of the free market system, *equal* access means that one does not hinder, or otherwise impede, individuals who have the funds and the interest to buy health care. One does not stop people from getting to the marketplace. And, in the two-tiered system, equal access means provision of a decent minimum to all. That decent minimum must be seen as satisfying a basic claim of beneficience, or as providing an allocation, that should it come our way, we would be minimally satisfied, or as being an allocation that we are willing to tolerate because it benefits the worst-off-class in society (and therefore we would not have rational grounds to protest if we come to be in that position) or as a way of dividing resources that produces the greatest amount of value over disvalue.

Those who are in favor of equal distribution of health care in principle as an interpretation of equal access must, however, hold that none of the above considerations in favor of a two-tiered system is the central consideration. They find something unfair and unjust, not simply unfortunate, or perhaps unfeeling, in a rich individual being able to purchase an exotic and expensive life-saving treatment that a poor individual cannot afford — even if there is the provision of a decent minimum via a two-tiered system, or even if it leads to the greatest amount of value over disvalue. How is one to interpret such a position? At best, it must be viewed as a position which holds that envy is a part of the character of the rational individual and that therefore jealousy is a rule of society (i.e., jealousy is when one is willing to take from another who has more, because he or she has more, even if that transfer means that one will end up with less oneself).[6] Note that this commitment to a principle of envy obtains if one wants to say that non-equal distribution is not simply distressing but indeed unfair. It would mean that one would hold that a rational individual would rather have all individuals, including himself or herself, have less health care even if disparities in health care distribution would lead to a higher minimal standard for all.

Notice, this characterization applies only to those who would on principle give an equal distribution of health care, not to those who hold that equal distribution of health care is a goal that one should, all else being equal, pursue. In particular, it is not a characterization of the argument by those who would want to hold in fact that an equal distribution would produce the greatest balance of value over disvalue. That is, one might argue that having a two-tiered system fails to put pressure upon the wealthy to provide an adequate health care system for those who are poor ([15], pp. 191–192). The latter would not be an argument in principle for equal health care.

And again, the problem is in understanding what would count as an equal distribution recurs. That is, is there to be a distribution of an equal amount of money to each individual for his or her health care? Or does it mean that everyone should be brought up to the same state of health? But of course that is impossible. Or does it mean that everyone should be brought up to a certain minimum state of health? Yet, even that is impossible. That natural lottery is most vicious when it comes to health. As already indicated, it is much easier to equalize wealth than to equalize health status. At best, it might mean making a reasonable *effort* to bring everyone up to the same level, or to a certain minimal level of health care — the attainment of a minimal level of health for all is obviously impossible unless one can abolish premature deaths, etc. Here of course it also becomes very important to decide what health is. And the difficulties here, as indicated, are legion, but surely one difficulty carrying the standards of that legion is the distinction between desires and needs. One has to be able to distinguish between health care desires and needs, or legitimate and pressing needs, versus mere desires or even frivolous desires.

IV. TESTING INTUITIONS

I hope at this point to have shown how difficult it is to develop a theoretical account of the questions of justice and fairness in the allocation of health care resources. I have not, however, wanted to suggest or imply that the difficulties are insurmountable. In any event, I do not think that we should despair about not having perfect answers. It is a sloppy world, and we will have to deal with sloppy answers; things worth doing well are usually worth doing poorly. That is, we may be able to sort our reasons for saying that some answers are better than others, even though absolute answers are not forthcoming. Towards that end I will sketch an artificial circumstance to aid in sorting our some of our intuitions and arguments on the matter of health care. One should imagine that he or she is part of a fairly large group of individuals stranded on an unowned, unclaimed island. Moreover, let us imagine that the landing party divides into four, freely formed groups, which decide unanimously to develop themselves into enduring political entities. They therefore survey the island into four sections held by all to be equally pleasant, draw straws, and divide it, and form four different communities, the A's, the B's, the C's, and the D's.

The A's decide against providing any health care whatsoever. They decide that life is short, and everyone of the A's agrees that they will live it up until they die, but that no health care will be provided. Towards that end, they ferment bananas and coconuts and other exotic fruits. The B's, however, provide preventive medicine, abortions on request and plastic surgery. The C's provide all health care they can and the D's do as well.

As things will have it, the D's possess a number of vicious recessive traits which begin to express themselves. One of their painful and debilitating genetic diseases can be treated only by the use of a rare fruit that grows on the A's part of the island. The A's, however, use it as an hallucinogen for their wild midnight revels. The D's also have a disease that can only be treated by an extract of bananas, but it would require restricting the diets of the A's, B's, and C's to provide enough of it to help the D's. However, such restriction would not lead to any severe hardship for the A's, the B's and the C's. Leaving aside special moral issues involved in prenatal diagnosis or negative eugenics, that is, considering just the living afflicted D's, does anyone have a duty to help them? Granted, it may be unfeeling and hard-hearted if no one helps, but is it unfair or unjust? Do the C's have a special obligation to the D's since they have also committed themselves wholeheartedly to health care? If the disease can be eliminated by prenatal diagnosis and abortion, do the B's have more of an obligation to the D's than do the A's? That is, does a commitment to a good or value for oneself commit one as well to help others to attain that good or value? Does such a common commitment make the failure to provide aid unfeeling or unkind? Does the difference in the view of the good life between the A's and the D's prevent one from saying that the A's are unfeeling?

I think that one can see that the way one will answer will depend on how one interprets the moral neutrality or non-neutrality of the natural lottery, and how one views entitlements. It will also depend upon one's concrete view of the good life. The more it is unlikely that there are strong moral arguments for one concrete view of the good moral life versus all others, the more ethics will offer procedural virtues. Thus one will be constrained to respect the freedom of others as a condition for a moral community not based on force. One will, however, have at best only weak arguments that a particular community should frame its version of the good life in a particular way. In fact, given that (1) the condition of respecting the free choice of others is integral to a community not based on force, and (2) the failure of moral arguments for particular concrete ways of life to be convincing, then one is led to the conclusion that rights to health care are to be created not discovered. There is, in short, no strong moral argument to allow one to choose among the A's, B's, or C's and D's.

In any event, the case of the island is artificial, and what one wants to ask is: What is it like to have a complex community without such a pure history? In particular, one may wish to ask how one can reconcile limited entitlements with interests in not being unfeeling, hard-hearted, or lacking in concern for the welfare of other persons. If one views the community as composed of individuals holding rights prior to that community and actually owning certain things, one can then see the community as a means of creating a large common resource to pursue common goods and values. Surely one of the goods that most communities would pursue would be adequate health care. Yet one might imagine communities that would choose not to do so. There would be nothing unjust in such a choice. One may in addition find communities that would provide somewhat bizarre means of distributing health care resources, yet since there is a fair agreement among humans about kinds of risks to avoid, one is likely to have at least some overlap among notions of rights to health care. The point is that communities create or invent, as much as discover, rights to health care and thus the sense of fairness for health care allocation systems. Such creations or inventions are dependent upon a culture and its particular history, its views of what virtues or excellences of character should prosper, and which should or can be given short shrift.

Thus one would have quite different notions of what health care allocations should be, if one has a fairly individualistic society, in contrast to a non-individualistic society. One might think of an individualistic society that provides a negative income tax plus a catastrophic health care plan (but allowed one to squander money received through the negative income tax, and thus not have funds left for health care, after which one would be denied health care), with a society that provided all its members with a certain minimum amount of health care. What would be acceptable or not-acceptable distributions will also turn on the importance a society vests in maximum efficiency. Thus in wartime most societies are willing to give the best health care to those who are likely to make the most important contributions to winning the war. The more one sees the general enterprise of one's society marked by such importunacy, the more it will be reasonable to allocate funds on such basis. Notice, what is important here is not so much how things get done, but rather that the people involved agree to the doing. This is because the community is properly seen to be derived through the free contributions of its members. But through those contributions enterprises become ineluctably socialized. As health care is increasingly subvened from common resources, it must increasingly be responsible to common wishes and pursue common goals. Uses of health care resources are unfair or unjust if they do not accord with why the resources were given. The exact character of justice in health care is more created or invented than discovered by the free individuals that constitute the community.

V. CONCLUSIONS

I have suggested that issues of the justice of health care allocations turn on how one distinguishes the unfair from the unfortunate, and that this is closely tied to how one assesses the significance of the natural lottery, which is closely associated with how we view the basis of entitlements to goods. Moreover, insofar as we as free individuals create a society, different views of moral excellence and character in the face of human finitude will compete for our attention. How we choose among them will determine which health care allocations are proper. In short, insofar as freedom functions as a side constraint, and therefore as long as the moral community is one based on respect of freedom not force, individuals will have the possibility of holding private entitlements. Moreover, they may or may not agree to a particular vision of an ideal moral distribution of goods, including goods for health care. In the absence of a shared common intuition of the good life, which intuition appears to be absent, the view of the good life will be created, not discovered. Any such creation will need to be pursued through:

(1) creating a line defining needs vs. desires, and
(2) deciding the extent to which one will treat unfortunate circumstances as if they were cases of unfairness.

In all circumstances one needs to anticipate with tolerance
(3) the subversive nature of freedom. For example, if one establishes an equal distribution of care and someone saves up on the side, can he or she buy more? Do they own the resources that they have saved up for the trade?

Exact answers to these questions must be created by a community. Absent shared views of the good through a common intellectual intuition, free communities will need to reconcile themselves to discussing in common and comparing different ways in which they might fashion the good life. Apart from respecting the freedom of those concerned in this discussion and in the common pursuit of a view of the good, there will be no univocally right or wrong answers. Rather, progress should be made via that grand propaganda of persuasion — the arts and literature that can in advance suggest to us how in different circumstances different sets of vices and virtues will alternatively thrive or fail to flourish.

Kennedy Institute of Ethics
Georgetown University
Washington, D.C.

NOTES

[1] A rather pure version of this position is offered by Robert M. Sade [14].
[2] See [5], [16], and [17]. The bill, "Health Care for All Americans Act" was introduced into Congress July, 1979.
[3] "... the women never upon any account eat with the men, but always by themselves. What can be the reason of so unusual a custom, it is hard to say; especially as they are a people in every other instance, fond of Society and much so of their Women. They were often asked the reason, but they never gave no other Answer, but that they did it because it was right, and expressed much dislike at the custom of Men and Women eating together of the same Victuals ... more than one-half of the better sort of the inhabitants have

entered into a resolution of enjoying free liberty in Love, without being troubled or disturbed by its consequences." "Both sexes express the most indecent ideas in conversation without the least emotion, and they delight in such conversation beyond any other. Chastity, indeed, is but little valued . . ." ([3], pp. 86–91).

[4] Compare John Rawls's list of the primary goods ([13], p. 62) with that of Ronald Green ([7], pp. 111–126).

[5] For a discussion of this point see ([12], p. 206).

[6] For an interesting discussion of the relationships between envy and jealousy, see ([12], pp. 239–242).

BIBLIOGRAPHY

1. Anon.: 1977, 'Pay Beds: Decree Nisi', *The Economist*, April 16, 22–23.
2. Anon.: 1978, 'Phasing Out Private Consulting Rooms', *London Times*, August 2.
3. Cook, J.: 1893, in Capt. W. S. L. Wharton (ed.), *Captain Cook's Journal 1768–71*, Elliot Stock, London.
4. Department of Health and Social Security: 1978, *Further Withdrawal of Pay Beds*, No. 78/145, London, May 10.
5. Detsky, A.: 1978, *Economic Foundations of National Health Policy*, Ballinger Publishing Company, Cambridge.
6. Fried, C.: 1976, 'Equality and Rights in Medical Care', *Hastings Center Report* 6, 29–34.
7. Green, R.: 1976, 'Health Care and Justice in Contract Theory Perspectives', in R. M. Veatch and R. Branson (eds.), *Ethics and Health Policy*, Ballinger Publishing Company, Cambridge, pp. 111–126.
8. Jonas, S. *et al.*: 1977, *Health Care Delivery in the United States*, Springer Publishing Company, New York.
9. MacIntyre, A.: 1980, 'Why is the Research for the Foundation of Ethics so Frustrating', in H. T. Engelhardt, Jr. and D. Callahan (eds.), *Knowing and Valuing: The Search for Common Roots*, The Hastings Center, Hastings-on-Hudson, New York, pp. 13–35.
10. Miles, R., Jr.: 1974, *The Department of H.E.W.*, Praeger Publishers, New York.
11. Myers, R. J.: 1970, *Medicare*, Richard D. Irwin, Inc., Homewood, Illinois.
12. Nozick, R.: 1974, *Anarchy, State and Utopia*, Basic Books, New York.
13. Rawls, J.: 1971, *A Theory of Justice*, Harvard University Press, Cambridge.
14. Sade, R. M.: 1971, 'Medical Care as a Right: A Refutation', *The New England Journal of Medicine* 285, 1288–1292.
15. Singer, P.: 1976, 'Freedom and Utilities in the Distribution of Health Care', in R. Veatch and R. Branson (eds.), *Ethics and Health Policy*, Ballinger Publishing Company, Cambridge, pp. 175–193.
16. Subcommittee on Health and Scientific Research: 1979, 'Health Care for All Americans Act', May 14.
17. U.S. Department of Health, Education and Welfare: 1976, *National Health Insurance Proposals: Provision of the Bills Introduced in the 94th Congress as of February 1976*, Social Security Administration, Office of Research and Statistics, DHEW Publication No. (SSA) 76-11920.

18. Veatch, R. M.: 1976, 'What Is a 'Just' Health Care Delivery?', in R. Veatch and R. Branson (eds.), *Ethics and Health Policy*, Ballinger Publishing Company, Cambridge, pp. 127–153.

JAMES F. CHILDRESS

PRIORITIES IN THE ALLOCATION
OF HEALTH CARE RESOURCES[1]

Students of biomedical ethics have traditionally concentrated on issues in the patient-physician relationship. In recent years, however, they have devoted increasing attention to issues in biomedical ethics in public policy. Public policies, defined as "whatever governments choose to do or not to do" ([7], p. 1), typically involve regulation (e.g., prohibition and control of an activity) and allocation and distribution of benefits (e.g., goods and services) and burdens (e.g., taxation). Issues in the allocation of resources for and within health care are among the most difficult from the standpoint of ethics. My task is to analyze some of these issues in allocation. Although I shall argue for positions at several points, my main intention is to provide a map of several major issues. I shall emphasize the *content* of public policies, not the *processes* by which they are formulated and implemented.[2] I shall also avoid some broad and important questions of social ethics regarding the structure of the health care system in the United States (e.g., whether the current mix of private and public is desirable). While some policies may imply changes in the structure of the health care system, I shall limit my attention and analysis to policies of allocation of resources for and within health care.

I want to concentrate on two major questions in the allocation of resources for and within health care:

1. What resources (time, energy, money, etc.) should be put into health care and into other social goods such as education, defense, eliminating poverty, and improving the environment?

2. Within the area of health care (once we have determined its budget), how much time, energy, money, etc., should we allocate for prevention and how much for rescue or crisis medicine?

These questions involve 'first-order determinations' or macroallocation decisions — how much of a good should be made available? They are distinguished from 'second-order determinations' or microallocation decisions — who should receive the available goods?[3] Although 'second-order determinations' presuppose 'first-order determinations', tensions and conflicts in the former may lead to a reassessment of the latter. Thus, tensions and conflicts in the allocation of scarce kidney dialysis and transplantation in part led to the federal government's decision to allocate funds to cover practically everyone who needs dialysis or transplantation.

I. HEALTH CARE VS. OTHER SOCIAL GOODS

Current evidence does not indicate that our great expenditures in health and medical care in, say, the last twenty years have brought us closer to health. In particular, our exotic technologies offer only marginal returns in reducing morbidity and premature death. The advances in health in the last century can be accounted for largely by improvements in living conditions, etc., rather than improvements in medical care. Therefore, the pursuit of some other social goods such as improving the environment and reducing poverty has beneficial effects on health.

If we accept the WHO definition of Health ("a state of complete physical, mental and social well-being") all social goods would relate directly or indirectly to health. Practically all allocation decisions would concern the aspects of health to be emphasized and the most effective and efficient means to their realization. But if we assume a narrower and more adequate definition of health (without developing the arguments for it at this point), we have to confront the conflict between health care, especially medical care, and some other social goods, not all of which serve as instruments to better health. For example, should hospitals always have priority over museums and opera houses? One philosopher, Antony Flew [8], has argued that "morally, so long as hospitals are needed, hospitals must always have priority over amusement parks" on the grounds that pain is not symmetrical with pleasure and that the prior or more fundamental duty is to alleviate pain. But it is not evident (a) that hospitals are primarily to alleviate pain, and (b) that they should always take priority over all other social goods that do not contribute to the aim of hospitals whether it is the alleviation of pain or some other goal. Health may be a condition for many values for individuals and the community, but it does not have finality or ultimacy. It is not true that when it comes to health, no amount is too much.

Paul Ramsey has said that he does not know how to go about resolving this first priority question — health care in relation to other social goods — because it is "almost, if not altogether incorrigible to moral reasoning" and "to rational determination" ([20], pp. 240, 268).[4] Ramsey suggests it is basically a *political* question, i.e., one to be resolved through political processes that can reflect the values, preferences and informal priorities of the society. Can one complain of *injustice* if a society puts more money into space programs or defense than health care? 'Wrong' priorities may not be unjust unless there are certain basic needs or rights that must be satisfied for justice to be realized.

Suppose we say that just policies give to each according to their needs and that medical care is a basic human need. Medical needs are unpredictable, random, and overwhelming, according to one line of argument, and society ought to be prepared to meet those needs. Such an argument would depend on a vision of human life that we cannot assume in our society — a vision that would rank needs and make health the most important one. Even if we distinguish needs and wants, demands for health care appear to be virtually without limit. And we have to ask how much we are willing to devote to the provision of medical care which, as we have seen, may not be all that important for health.

Another line of argument is that there is what Charles Fried calls a right to a decent minimum of health/medical care [11], and that society should make sure that there is enough in the budget to meet this need. An argument for equal access to medical care does not necessarily imply a minimum for individuals and thus for the health budget. It may only mean equal access to what is available. And what is available may be meager. But a right to a *decent minimum* establishes a base for individual medical care and, consequently, for the health budget. It would provide a standard for determining the minimum amount for the health care budget. But, again, in the absence of a shared vision of humanity, we may have to resort to the political process to define the decent minimum for individuals.

In short, to determine how to allocate resources for health care in relation to other social goods, we need to resolve several matters; the definition of health, its value, its causes, whether there is a right to a decent minimum of health/medical care, and what that minimum is.

II. WITHIN THE HEALTH CARE BUDGET, HOW MUCH SHOULD WE ALLOCATE FOR *PREVENTION* AND HOW MUCH FOR *RESCUE* OR *CRISIS MEDICINE*?

Although this conflict between prevention and crisis or rescue medicine rarely emerges in a clear and manageable form in debates about public policy, it is present and needs identification and analysis so that we can appreciate the 'trade-offs' that we frequently, if unwittingly, make. *Prevention* includes strengthening individuals (e.g., through vaccines), changing the environment, and altering behavioral patterns and lifestyles. Many recent commentators insist that the most effective and efficient way to improve the nation's health is through *prevention*, since our current emphasis on rescue medicine now produces only marginal returns. DHEW's *Forward Plan for Health, Fiscal Years 1977–1981*, holds: "Only by preventing disease from occurring, rather than treating it later, can we hope to achieve any major improvement in the Nation's health" ([24], p. 15; cf. [17]).

This recommendation is at odds with our current macroallocation policies. For example, in 1976 expenditures for health amounted to 11.4% of the federal budget. When that amount (42.4 billion dollars) is assigned to four major determinants of health (human biology, lifestyle, environment, and the health care system), the results are striking: 91% went to the health care system, while 3% went to human biology, 1% to lifestyle, and 5% to environment [16].

Nevertheless, the evidence for the effectiveness and efficiency of a preventive strategy to reduce morbidity and premature mortality by concentrating on human biology, lifestyle, and environment is by no means conclusive. The appropriate mix of preventive and rescue strategies will depend in part on the state of knowledge of causal links. Despite the dramatic example of polio, some other conditions such as renal failure are not the result of a single disease or factor. As a consequence, prevention appears to be only a remote possibility. In some cases prevention (which might involve extensive and expensive screening in order to identify a few persons at risk) may be less cost-effective than therapy after the disease has manifested itself. Our uncertainty in these areas may be an argument for increased research.

Even if a preventive program (at least in certain areas) would be more effective and efficient, its implementation would not be free of moral, social, and political difficulties. Effectiveness and efficiency, or utility, are not the only relevant standards for evaluating policies of allocation of resources within health care (which is defined broadly and includes more than medical care). I shall concentrate on (1) the symbolic value of rescue efforts (what is symbolized about both the victim and the rescuer), (2) the duty of compensatory justice generated by the principle of fairness, (3) the principle of equal access, and (4) the principle of liberty.

(1) Our society often favors rescue or crisis intervention over prevention because of our putative preference for known, identified lives over statistical lives. The phrase 'statistical lives' refers to unknown persons in possible future peril ([22], cf. [25]). They may be alive now, but we do not know which ones of them will be in future peril. Mining companies often are willing to spend vast sums of money to try to rescue trapped miners when

they will spend little to develop ways to prevent such disasters even if they could save more lives *statistically* in the long run. And at 240 million dollars a year, we could save 240 lives of workers at coke ovens. We do not know which ones will be saved, and we are not likely as a society to respond enthusiastically to this expenditure. But we cannot ignore statistical lives in a complex, interdependent society, particularly from the standpoint of public policy.

The principle of equality is not violated by including statistical lives in public policy deliberations. It is not merely a matter of sacrificing present persons for future persons, for there are two different distinctions to consider: On the one hand, the distinction between *known* and *unknown* persons; on the other hand, the distinction between *present* and *future* peril ([10], pp. 224 f.; Fried's argument is important for much of this section of the paper). Existing and known persons who may not be in present danger may be in future danger if certain preventive measures are not taken. Thus, preventive measures may aid existing persons who are at risk and not merely future persons.

Following Max Weber, it is possible to distinguish between 'goal-rational' (*zweckrational*) and 'value-rational' (*wertrational*) ([26], p. 1).[5] Conduct that is 'goal-rational' involves instrumental rationality — reasoning about means in relation to ends. Conduct that is 'value-rational' involves matters of value, virtue, character, and identity that are not easily reduced to ends, effects, or even rules of right conduct. There is thus an important distinction between *realizing* a goal and *expressing* a value, attitude, or virtue. In the context of debates about prevention and rescue intervention, the 'goal-rational' approach concentrates on effectiveness and efficiency in statistical terms, while the 'value-rational' approach focuses on the values, attitudes, and virtues that policies express.

This distinction between 'value-rational' and 'goal-rational' conduct may illuminate the 1972 congressional decision to make funds available for almost everyone who needs renal dialysis or transplantation. This decision followed widespread publicity in the media about particular individuals who were dying of renal failure. One patient was even dialyzed before the House Ways and Means Committee. Now we have a program that costs over 1 billion dollars each year. Some argue that this decision was an attempt to preserve society's cherished myth that it will not sacrifice individual lives in order to save money. Of course, we make those sacrifices all the time (e.g., when we fail to pass and enforce some safety measures). But society's myth is not as threatened when the sacrificed lives are statistical rather than identified.

Hence, decisions to try to rescue identified individuals have symbolic value. It has been said that the Universalists believe that God is too good to damn men, and the Unitarians believe that man is too good to be damned. Similarly, the symbolic value argument suggests that rescue attempts show that individuals are 'priceless' and that society is 'too good' to let them die without great efforts to save them. This is society's myth. And, so the argument goes, when Congress acted to cover the costs of renal dialysis, it acted in part to preserve this myth, for the "specific individuals who would have died in the absence of the government program were known" ([30], pp. 447–48).[6] They were identified lives. The policy was 'value-rational', if not 'goal-rational'.

Insofar as this argument focuses on the way rescue interventions symbolize the value of the victims, it encounters difficulties as a basis for allocative decisions. Consider two possibilities: (a) we keep the same total lifesaving budget but withdraw resources from *preventive* efforts in order to put them in *rescue* efforts so that we can gain the symbolic value of crisis interventions. As Charles Fried argues, "surely it is odd to symbolize our concern for human life by actually doing less than we might to save it," by saving fewer lives than we might in a maximizing strategy. (b) Another possibility is to keep the same prevention budget but to take resources from other areas of the larger budget (cf. my discussion of I) so that we can increase crisis interventions. In this case, however, "we symbolize our concern for human life by spending more on human life than in fact it is worth" ([10], p. 217).

The symbolic value argument focuses not only on the symbolic value of the victims, but also on what is symbolized about the agents: the virtue and character of the society and its members, what is sometimes called 'agent-morality'. Conduct that is 'value-rational', in contrast to 'goal-rational', may thus be based on an answer to the questions 'who are we?' or 'who shall we be?' Allocation policies may be thought of as ways for the society to define and to express its sense of itself, its values, and its integrity. From this standpoint, it is possible to argue that rescue efforts should be important, if not dominant, in allocative decisions. As Lawrence Becker notes, "we have (rationally defensible) worries about the sort of moral character represented by people who propose to stand pat and let present victims die for the sake of future possibilities. One who can fail to respond to the call for help is not quite the same sort of character as one who can maximize prevention" ([1], p. 118).[7] Although 'agent-morality' has a strong appeal, and no doubt influences many of our decisions, it is difficult to determine how much weight it should have in our policy deliberations.

(2) In some situations the principle of fairness can generate a duty of *compensatory justice* that assigns priority to identified lives in present danger. In such situations society has an obligation to try to rescue individuals, even though it departs from a strategy that would save more lives in the long run. These situations involve some unfairness because of an inequality in the distribution of risks which the society has assigned, encouraged, or tolerated. Suppose that a person who has worked in a coal mine under terrible conditions that should have been corrected is trapped in a cave-in or comes down with a disease related to his working conditions. Fairness requires greater expenditures and efforts for him in order to equalize his risks. Likewise, if society fails to correct certain environmental conditions that may interact with a genetic predisposition to cause some diseases, it may have a duty of compensatory justice. A policy of compensatory justice in health-related matters, of course, faces numerous practical difficulties, such as identifying causation. But it is important to underline the point that the principle of fairness can generate a duty of compensatory justice that sets limits on utility and efficiency in some situations. Those who argue that individuals voluntarily choose to bear risks and thus waive any claim to compensatory justice need to show that the individuals in question really understand the risks and voluntarily assume them (e.g., do the workers in an asbestos factory have an opportunity to find other employment, to relocate, etc.?). But even the voluntary assumption of risks should not always be construed as a waiver of a claim to compensatory justice when the person is injured (e.g., research-related injuries).[8]

(3) If we include statistical lives in a preventive strategy, while allowing for compensatory justice, we still face difficult questions about *distributive justice*, particularly the principle of equality or equal access. It is not enough to maximize aggregate health benefits. Our policies should also meet the test of distributive justice, which is, indeed, presupposed by compensatory or corrective justice. For example, consider a program to reduce hypertension (high blood pressure) which affects approximately 24 million Americans and poses the risk of cardiovascular disease. In order to reduce the morbidity and mortality from cardiovascular disease, Weinstein and Stason propose an anti-hypertensive program on the basis of cost-effectiveness analysis. They recommend the intensive management of known hypertensives instead of public screening efforts. As they recognize, this proposal appears to be disadvantageous to the poor who would probably be unaware of their hypertensive condition because of their limited access to medical care. This inequity might be diminished by Weinstein and Stason's proposal to give target

screening in black communities priority over community-wide screening since hypertension is more common among blacks than among whites, and by selective screening in low-income communities.[9]

The formal principle of equality, or justice, is 'treat similar cases in a similar way.' Of course, such a formulation does not indicate the relevant similarities. When we discuss equal access to medical care, we most often consider medical need, in contrast to geography, finances, etc., as the relevant similarity that justifies similar treatment. Suppose we decide that an effective and efficient strategy to improve the nation's health will not permit us to do all we could do in rescue or crisis medicine. A decision to forego the development of some technology or therapy may condemn some patients to continued ill health and perhaps to death. Can such a policy be justified?

One possible approach that still respects the formal principle of justice and excludes arbitrary distinctions (e.g., geography and finances) between patients would exclude 'entire classes of cases from a priority list'. According to Gene Outka, it is more "just to discriminate by virtue of categories of illness, for example, rather than between rich ill and poor ill" ([19], p. 24). Society could decide not to allocate much of its budget for treatment of certain diseases that are rare and noncommunicable, involve excessive costs, and have little prospect for rehabilitation. The relevant similarity under conditions of scarcity would not be medical need but category of illness. While certain forms of treatment would not be developed and distributed for some categories of illness, care would be provided. Patients would not be abandoned.

For allocation of funds for research into the prevention and treatment of certain diseases, it is important to consider such criteria as the pain and suffering various diseases involve, their costs, and the ages in life when they are likely to occur [23]. Applying some criteria such as pain and suffering, we might decide to concentrate less on killer diseases such as some forms of cancer and more on disabling diseases such as arthritis. Thus, says Franz Ingelfinger, national health expenditures would reflect the same values that individuals express: "it is more important to live a certain way than to die a certain way" [13].

One danger of such an approach to priorities among diseases should be noted: a decision about which diseases to treat may in fact conceal a decision about which population groups to treat since some diseases may be more common among some groups [15]. Injustice may result from these priorities.

(4) The principle of *liberty* also poses some moral, social, and political

difficulties for an effective and efficient preventive strategy. Mounting evidence indicates that a key determinant of an individual's health is his or her *lifestyle*. Leon Kass argues that health care is the individual's duty and responsibility, not a right [14]. And Lester Breslow offers seven rules for good health that are shockingly similar to what our mothers always told us! The rules, based on epidemiological evidence, are: don't smoke, get seven hours sleep each night, eat breakfast, keep your weight down, drink moderately, exercise daily, and don't eat between meals. At age 45, one who has lived by 6 of these rules has a life expectancy eleven years longer than someone who has followed fewer than 4 ([2, 3]).

On the one hand, we have increasing evidence that individual behavioral patterns and lifestyles contribute to ill health and premature mortality. On the other hand, we have the liberal tradition that views lifestyles as matters of private choice, not to be interfered with except under certain conditions (e.g., harm to others).

Each person has what Charles Fried calls a *life plan* consisting of aims, ends, values, etc. That life plan also includes a *risk budget*, for we are willing to run certain risks to our health and survival in order to realize some ends ([10], Chap. 10–12). We do not sacrifice all our other ends merely to survive or be healthy. Our willingness to run the risk of death and ill health for success, friendship, and religious convictions discloses the value of those ends for us and gives our lives their style. Within our moral, social, and political tradition, the principle of liberty sets a presumption against governmental interference in matters of lifestyle and voluntary risk-taking. But that presumption may be overridden under some conditions. Let me make a few points about those conditions which are similar to just war criteria:

(1) An important goal is required to override liberty. One goal is *paternalistic*: to protect a person even when his actions do not harm anyone else. This is rarely a sufficient justification for interference with liberty. Usually we require that restrictions of liberty be based, at least in part, on the threat of harm to others or to the society (e.g., compulsory vaccinations). We have difficulties with purely paternalistic arguments in requiring seatbelts, etc. Another goal might be to *protect the financial resources of the community*. If we get national health insurance, we can expect increased pressure to interfere with individual liberty. Why? People simply will not want to have their premiums or taxes increased to pay for the *avoidable afflictions* of others. They will charge that such burdens are unfair.

(2) To override the presumption against interfering with liberty, we need strong evidence that the behavior or lifestyle in question really

contributes to ill health. This second standard must be underlined because there is a tendency to use a term like 'health' which has the ring of objectivity to impose other values without articulating and defending those values. In Camus's novel, *The Plague*, a doctor and a priest fighting the plague in an Algerian city engage in the following conversation. The priest says to the doctor: "I see that you too are working for the salvation of mankind." "That is not quite correct," the doctor replies, "Salvation is too big a word for me. I am working first of all for man's health". The danger is that 'salvation' or 'morality' or a certain style of life will be enforced in the name of health although it has little to do with health and more to do with the legislation of morality.

(3) Another condition for overriding the presumption against interfering with liberty is that the interference be the *last resort*. Other measures short of interference such as changing the environment should be pursued and sometimes continued even when they are less effective and more costly than measures that restrict liberty.

(4) A fourth condition is a reasonable assurance that the restriction will have the desired result as well as a net balance of good over evil.

(5) Even when we override the principle of liberty, it still has an impact. It requires that we use the least restrictive and coercive means to reduce risk-taking. For example, information, advice, education, deception, incentives, manipulation, behavior modification, and coercion do not equally infringe liberty. In addition to choosing the least restrictive and coercive means, we should evaluate the means on other moral grounds.

These considerations and others come into play when we try to determine whether (and which) incursions into personal liberty are justified. To allocate resources to prevention rather than to rescue is not a simple matter, for successful prevention may infringe autonomy and other moral principles.[10]

University of Virginia
Charlottesville, Virginia

NOTES

[1] This is a revised version of an article with the same title that appeared in *Soundings* LXII (Fall, 1979), 256–274. The journal *Soundings* holds the original copyright.
[2] For an examination of *processes* in 'tragic choices', see [4], particularly their emphasis on openness and honesty in processes of allocation.
[3] For the distinction between 'first-order' and 'second-order' determinations, see [4].

[4] By contrast, Ronald Green [12] argues that the question of how much the society should spend on health should be handled in terms of general moral principles rather than left to the political process. Within contract theory, he contends, "members of the original position as the architects of the basic social system want to set some upper and lower limits on the availability of health services." Green's argument depends on a modification of Rawls's theory of primary goods to include health under the "social primary goods" rather than under the "natural primary goods" whose distribution is only indirectly affected by the social structure.

[5] The translators use the term 'purpose-rational' for '*zweckrational*'.

[6] Contrast [21].

[7] See also [18] and [9].

[8] For a fuller discussion of compensatory justice, see [5] and ([10], p. 220).

[9] See [27], [28] and [29]. Recent data indicate that "the stroke death rate, when adjusted for age changes in the population, has declined 36.1 percent since 1962, with more than two-thirds of that drop occurring since 1972, the year a major, continuing national campaign was begun to identify and treat those suffering from high blood pressure" ([6], pp. A1, A9).

[10] The formulation and development of the ideas and arguments in this essay benefitted greatly from discussions at many colleges, universities, medical schools, and other institutions. Among the most fruitful discussions were those at the University of Tennessee (Knoxville), The Center for Disease Control, The National Institutes of Health, the School of Public Health of the University of North Carolina (Chapel Hill), The Albert Einstein College of Medicine of Yeshiva University, The Institute of Religion at the Texas Medical Center, Princeton Theological Seminary, and Indiana University, which subsequently reproduced and circulated one version of this paper. I am grateful to the persons at these and other institutions who sponsored my lectures and who contributed incisive criticisms and suggestions.

BIBLIOGRAPHY

1. Becker, L. C.: 1975, 'The Neglect of Virtue', *Ethics* 85, 110–122.
2. Belloc, N. B.: 1973, 'Relationship of Health Practices and Morality', *Preventive Medicine* 2, 67–81.
3. Belloc, N. B. and Breslow, L.: 1972, 'Relationship of Physical Status and Health Practices', *Preventive Medicine* 1, 409–421.
4. Calabresi, G. and Bobbitt, P.: 1975, *Tragic Choices*, W. W. Norton & Co., New York.
5. Childress, J. F.: 1976, 'Compensating Injured Research Subjects: The Moral Argument', *Hastings Center Report* 6, (December), 21–27.
6. Colen, B. D.: 1979, 'Deaths Caused by Strokes Fall Sharply in U. S.', *The Washington Post*, February 23, pp. A1, A9.
7. Dye, T. R.: 1975, *Understanding Public Policy*, 2nd ed., Prentice-Hall, Inc., Englewood Cliffs.
8. Flew, A.: 1967, 'Ends and Means', in P. Edwards (ed.), *The Encyclopedia of Philosophy*, Vol. 2, Macmillan Publishing Co., and The Free Press, New York, p. 510.

9. Freedman, B.: 1977, 'The Case for Medical Care, Inefficient or Not', *The Hastings Center Report* **7** (April), 31–39.
10. Fried, C.: 1970, *An Anatomy of Values: Problems of Personal and Social Choice*, Harvard University Press, Cambridge.
11. Fried, C.: 1976, 'Equality and Rights in Medical Care', in J. G. Perpich (ed.), *Implications of Guaranteeing Medical Care*, National Academy of Sciences, Institute of Medicine, Washington, D. C., pp. 3–14.
12. Green, R.: 1976, 'Health Care and Justice in Contract Theory Perspective', in R. M. Veatch and R. Branson (eds.), *Ethics and Health Policy*, Ballinger Publishing Co., Cambridge, pp. 111–126.
13. Ingelfinger, F. J.: 1972, 'Haves and Have-Nots in the World of Disease', *New England Journal of Medicine* **287** (December 7), 1199.
14. Kass, L.: 1975, 'Regarding the End of Medicine and the Pursuit of Health', *The Public Interest* **40** (Summer), 11–42.
15. Katz, J. and Capron, A. M.: 1975, *Catastrophic Diseases: Who Decides What?*, Russell Sage Foundation, New York.
16. Koleda, M. S. et al.: 1977, *The Federal Health Dollar: 1969–1976*, National Planning Association, Center for Health Policy Studies, Washington, D. C.
17. LaLonde, M.: 1974, *A New Perspective on the Health of Canadians: A Working Document*, The Government of Canada, Ottawa.
18. LaRue, L. H.: 1972, 'A Comment on Fried, Summers, and The Value of Life', *Cornell Law Review* **57**, 621–631.
19. Outka, G.: 1974, 'Social Justice and Equal Access to Health Care', *Journal of Religious Ethics* **2** (Spring), 11–32.
20. Ramsey, P.: 1970, *The Patient as Person*, Yale University Press, New Haven.
21. Rettig, R. A.: 1976, 'Valuing Lives: The Policy Debate on Patient Care Financing for Victims of End-Stage Renal Disease', *The Rand Paper Series*, The Rand Corporation, Santa Monica.
22. Schelling, T.: 1966, 'The Life You Save May Be Your Own', in S. B. Chase, Jr. (ed.), *Problems in Public Expenditure Analysis*, The Brookings Institution, Washington, D.C., pp. 127–166.
23. Taeuber, C.: 1976, 'If Nobody Died of Cancer ... ', *The Kennedy Institute Quarterly Report* **2**, 6–9.
24. United States Department of Health, Education and Welfare, Public Health Service: 1976, *Forward Plan for Health: FY 1977–1981*, DHEW Publication No. (OS) 76-50024.
25. Weaver, W.: 1961, 'Statistical Morality', *Christianity and Crisis* **20**, 210–213.
26. Weber, M.: 1967, *Max Weber on Law in Economy and Society*, in M. Rheinstein (ed.), (transl. by E. Shils and M. Rheinstein), Simon and Schuster, New York.
27. Weinstein, M. C. and Stason, W. B.: 1977, 'Allocating Resources: The Case of Hypertension', *Hastings Center Report* **7** (October), 24–29.
28. Weinstein, M. C. and Stason, W. B.: 1977, 'Allocation of Resources to Manage Hypertension', *New England Journal of Medicine* **296**, 732–739.
29. Weinstein, M. C. and Stason, W. B.: 1976, *Hypertension: A Policy Perspective*, Harvard University Press, Cambridge.
30. Zeckhauser, R.: 1975, 'Procedures for Valuing Lives', *Public Policy* **23**, 419–464.

BARUCH BRODY

HEALTH CARE FOR THE HAVES AND HAVE NOTS: TOWARD A JUST BASIS OF DISTRIBUTION

Health care seems to be a basic need of all individuals. It goes with such other basic needs as food, clothing, and shelter. This may help explain our national commitment (in such programs as Medicaid and Medicare) to providing adequate health care for the indigent. Despite this commitment, and the great costs incurred in trying to meet it, the problem of providing adequate health care for the medically indigent (the have-nots) at an affordable cost to the taxpayers (the haves) seems to be unsolved. Some see the problem as being one of excessive costs. Some see the problem as being one of inadequate coverage; too many Americans who are medically indigent are not assured of adequate medical care. Some see the problem as being one of quality control; the medically indigent are receiving second-rate medical care. In truth, all of these problems are present. We are providing inadequate health care to an insufficient number of people at an excessive and rapidly increasing cost.

All of these issues have led many to reassess this national commitment, to ask whether we are doing too much or too little. This essay is part of this process of reassessment. Its goal is to seek out the roots of our national commitment. The hope is that such an analysis will lead us to structure our programs so that we may at least alleviate some of these problems.

In the first section of this essay, I will argue that the programs in question have to be viewed in the context of programs of redistribution aimed at promoting greater equality. In the second section, I will outline a theory to justify such redistributions. In the final section, I will develop some of the policy implications of adopting this justificatory theory.

I

I would like to begin by making two crucial points about programs of providing medical care for the indigent, viz., that they are programs of distributive justice and that they could in theory be made unnecessary by the direct provision of adequate funds to the indigent. Let us look at each of them separately.

The first point is that these are programs of taking funds from better-off taxpayers to provide health care for the indigent on the grounds that justice demands this equalization of capacity to obtain basic needs. Several implications follow from this point: (a) the problem of providing health care should not be separated from other problems of distributive justice; it should not be separated from other problems concerning the taking of funds from better-off taxpayers to provide other basic needs to the indigent. Some of these other basic needs (e.g., food and housing) are obviously related to health care, while others (e.g., education) are less so. Still, all must be supplied through funds raised from the same taxpayers, so we need to consider all of these needs together; (b) since the justification for these programs is that they are providing a basic need to those who are in need, the programs should apply to all who are medically indigent, to all who cannot pay for their medical care. There is no reason why certain indigent (e.g., the aged, or children with only one parent living at home) should be covered while others are not.

This point and its implications have been challenged implicitly by Marc Plattner in an important recent essay [5]. Plattner wants to distinguish legitimate social welfare programs (among which are Medicare and Medicaid) from questionable redistributive programs:

Social insurance and assistance to the needy can be regarded as legitimate functions of the public sphere, properly supported by public revenues. The obligation of citizens to pay taxes to finance the legitimate expenses of government has never been doubted by the liberal tradition. Financing public-welfare expenditures with tax dollars in no way conflicts with the notion that people have a right to what they earn, and that their own property is genuinely private. But as social programs grow larger and more complex, it is all too easy to make the mistake of regarding redistribution as a logical extension of – or even simply a way of rationalizing – the welfare spending of the liberal state. This is an error to which economists are particularly prone, given their penchant for focusing on economic effects (i.e., how much is being transferred to whom) rather than political principles (i.e., on what grounds the money is being transferred) ([5], pp. 47–48).

There are two points I want to make in response. To begin with, Plattner here assumes, and elsewhere states explicitly, that redistributive programs, unlike programs of social insurance and of assistance to the needy, necessarily conflict with the notions that people have a right to what they earn and that there is such a thing as legitimate genuinely private property. In the next section I shall try to suggest just the opposite, viz., that legitimate genuinely private property and the right to what one earns presupposes some redistributive measures. Secondly, it is hard to see what is the distinction which Plattner is drawing. He concedes that programs of assistance to the needy (such as Medicaid) are in fact redistributive, and claims that the only difference between his social welfare programs and redistributive programs are their justifications. But what could be the justification for such need-based programs except some notion of just redistribution? He says that, "one can easily accept the principle of public insurance on the one hand and public charity or relief on the other without acknowledging the propriety of governmental efforts to promote the goal of greater equality of incomes" ([5], p. 29). But what could be the justification for public charity except some notion of just redistribution? To be sure, one can advocate need-based programs (public charity) without advocating full-scale redistribution; but that is not the issue. One must at least be claiming that justice demands at least that much increase in equality as is produced by one's program, and to that extent, such programs are redistributive programs.

The second major point which I wish to make in this section is what programs for providing medical care for the indigent would be made unnecessary if we simply redistributed sufficient funds to the indigent so that they could pay for their basic needs including health care. I am not now saying either that distributive justice requires such a redistribution of wealth or that such a redistribution is better than the in-kind provision of health care. All that I want to claim for now is that if we did have such a redistributive program, the macro-problem of justice in the provision of health care would disappear.

The following case, it seems to me, both explains this second point and justifies it: suppose that we had distributed to the indigent sufficient funds so that they could pay for their basic needs including health care. Suppose, moreover, that one such recipient then used this aid for other purposes, became ill and was unable to pay for health care, and came to ask for more help. Would we be compelled as a matter of justice (as opposed to private mercy) to aid that person? I think not, and for two reasons. Aiding that person would be requiring those who had already paid for the just distribution to pay for it again, and that seems unfair. Moreover, an autonomous decision-maker can hardly claim a right to protection from the unfortunate consequences of his own freely chosen errors.

In fact, our own society has, for the most part, chosen a different technique for providing health care to the indigent. We have provided them with a medical insurance policy. One of the crucial policy questions with which we will have to grapple in Section III is whether this was a wise choice.

II

Let us now turn to the question as to whether justice demands the provision of health care to the medically indigent. I shall treat that question as equivalent to the question as to whether the medically indigent have a right to health care.

There are those who derive the existence of the right to health care from the assumption that we all equally need that care. Their argument runs roughly like this: we all equally need health care; therefore, we all have an equal right to health care; but the operations of the market result in some not having that care; therefore, society must redistribute wealth so that health care (or at least the means to purchase it) is available to everyone. I have never seen the force of this argument. After all, even if we do equally need health care, an equal right to health care follows only if one assumes that the need generates a right. This crucial assumption is left totally unsupported by the argument.

There are those who derive the existence of the right to health care from the assumption that the existence of transfers adequate to insure health care maximizes general welfare. Their argument runs roughly like this: any reasonable assumptions about the diminishing marginal utility of money, even when combined with assumptions about the need of monetary incentives to maximize production, lead to the conclusion that utility will be increased by the distribution of funds to purchase health care to the indigent. Therefore, the indigent have a right to this redistribution. I have never seen the force of this argument either. Its crucial difficulty is its move from maximizing utility to rights. Despite the ingenuity of Mill and those who have followed his theory of rights, it would seem as though rights are moral considerations independent of considerations of general utility, in that policy A's leading to a maximization of utility does not entail that there is a right to the carrying-out of policy A.

There are those who derive the existence of the right to health care from the existence of other rights. Their argument runs roughly like this: the effective enjoyment of these other rights (e.g., political rights) presupposes the possession of a certain basic set of goods. Since everyone has these other rights, it follows that everyone also has a right to the possession of this basic set of goods. But not everyone possesses these basic goods. Therefore, society must distribute wealth so that these basic goods (or at least the means to purchase them) are possessed by everyone. Health care is one of these goods, so society must provide it for the indigent. Again, I have never seen the force of this argument. Its crucial difficulty is the assumption that if everyone has certain rights, it follows that they also have the right to what is required for the effective enjoyment of these rights, an assumption left unsupported by the argument.

I do, nevertheless, think that an argument can be mounted from the theory of justice for a program of redistribution that certainly has implications for the right to health care. Let me now sketch that theory of justice and redistribution, a theory on which I am currently working, and which I have tentatively called quasi-libertarianism.[1] It is best understood by first considering libertarianism.

The popular image of the libertarian is that of someone who is opposed to extensive government activity, someone who believes that the only legitimate functions of the state are the watch-dog functions of protecting us against force, theft, and fraud. This image is, as far as it goes, correct, but it leaves out what is truly fundamental in the libertarian picture of man and society.

To the libertarian, the fundamental fact about man is that he is a rational agent who often chooses to act in certain ways, from among the alternatives open to him, because of his beliefs that these actions are either intrinsically best or best in leading to what he does desire intrinsically. Not all of these choices are deliberative, but many are. To the libertarian, man's freedom to act this way without being restrained or coerced by his fellow human beings is of fundamental importance and one of the fundamental human rights is the right not to be constrained or coerced from so acting by others. It is this which provides a foundation for the libertarian's opposition to most of the activities of the welfare state. The welfare state, in pursuing its goals of maximizing general welfare, of redistributing wealth, etc., constantly passes laws and regulations which are coercive in that they impose substantial penalties upon our acting in certain ways.

In theory, these basic libertarian views are independent of any views about the institution of private property and of any views about an individual's right to property. In fact, however, most writers in this tradition from Locke on have held that besides our right to life and bodily integrity and our right of freedom from coercion, we also have a right to the 'fruit of our labor', to the value produced by our labor. The contemporary libertarian [2] puts that in the form of a right to the value of any unowned commodity upon which we have worked (initial property rights) or to the value that others freely agree to give to us either in exchange for our giving them something else, e.g., our labor-power for a certain period of time, or gratuitously (transferred property rights). Because libertarians hold these additional views about property and the right to property, they have a second objection to the welfare state. The welfare state, in order to finance its programs of maximizing general welfare, of redistributing wealth, etc., takes the property to which we have a right, and this is another reason for its illegitimacy (libertarians, in fact, have to do a lot of work to explain how even the minimal state is to finance its activities).

Naturally, a lot more has to be said by way of explaining and justifying all of these claims, and I cannot do this in the context of this paper. What I do want to say is that (a) it seems to me that libertarianism, understood this way, is an important doctrine embodying certain correct perceptions about the relation between man and the state but that (b) it requires modification. Let me elaborate upon one crucial modification, for it leads to the quasi-libertarianism and the theory of distributive justice which I wish to advocate.

The modification which I would introduce into the libertarian account has to do with the question of the distribution of wealth. From the strict libertarian point of view, the only wealth to which an individual is entitled is that wealth which he has acquired by his labor on unowned commodities or that wealth which has been freely transferred to him by others already entitled to it. There is no special pattern of distribution about which individuals can claim that they have a right to get what they would have under this pattern. I would suggest, however, that there are libertarian grounds for modifying this strict conclusion.

The crucial idea behind the libertarian theory of property, going back to Locke ([1], chap. 5), is that initial property rights are ultimately grounded in entitlements to the value produced by labor, and that all property rights arise from these initial entitlements. But some wealth which exists is simply the initial value of natural resources, and neither Locke nor anyone else in the libertarian tradition has ever really explained why anyone should have an entitlement to that wealth. My argument[2] suggests that the simultaneous existence both of property rights and rights to redistributive welfare arises out of the difference between labor-created wealth and the wealth which is the value of natural resources. To see how this works, let us imagine an initial position of a social contract. All those forming the contract recognize the existence of equal libertarian rights not to be prevented or coercively threatened from using the natural resources of the earth. They also recognize that allowing exclusive property rights, which must be over natural resources as well as labor-created values, is economically efficient. What I claim is that such people would agree (a) to allow for the formation of exclusive property rights over natural resources as well as added values, (b) to compensate those who would lose the rights to use the natural resources assigned to property values, and (c) to provide that compensation in the form of socially-recognized welfare rights, socially-recognized rights to a minimum level of support.

I have found it useful, in thinking out the implications of this approach, to imagine the terms of the social contract running something like this: the natural resources of the earth are leased to those who develop them, or to those to whom they transfer those leases. In return, they owe a rental to everyone. That rental is collected as taxes and paid into a social insurance fund which covers everyone equally. The social insurance fund insures us against destitution, and it pays payments to those who are destitute. Like all insurance funds, while all are equally covered, not all receive payouts much less equal payouts.

Several crucial points follow even on the basis of this sketch: (a) most property rights are rights over a mixture of natural resources and added values. They can exist, therefore, only if the conditions agreed to in the initial social contract are met. Consequently, legitimate private property rights presuppose some redistribution of wealth, and Plattner is wrong in supposing that the two conflict; (b) the theory of social justice which is the heart of quasi-libertarianism leads to a theory of redistribution of wealth and not to special rights to any particular basic need. This provides a theoretical justification for the claim made in the last section that the problem of health care should not be separated from other problems of distributive justice. It should not be separated because what justice gives rise to is a right to certain redistributions. The right to health care is at best derivative from that more general distributive right; (c) since the fundamental right generated by this theory of justice is the right of the indigent to payments from the social insurance fund, all of the demands of justice would be satisfied by such a redistribution. So our theory provides a theoretical justification for a generalized version of our other main claim in the last section, viz., that there would be no macro-problem of justice in the provision of health care if there were in effect an adequate general system of redistribution.

III

Suppose that one were to agree that the demands for just redistribution are properly grounded in the sorts of considerations that were sketched in the last section. What would follow for the question of medical care for the indigent, viewed, as it should be, as a problem of distributive justice?

In order to answer this question, two corollaries of the theory must first be mentioned, corollaries having to do with the level of payments from the social insurance fund. In general, those who have argued that the indigent have rights either to cash payments or to in-kind benefits have not adequately considered the question of the level of payments or benefits demanded by justice, and yet that is obviously a crucial question. On our account, the level is determined by two factors — the amount paid to the insurance fund by the taxpayers, and the number of recipients drawing benefits. The amount paid to the insurance fund should be a fair rental for the natural resources, and the number of recipients will depend upon eligibility requirements. Both of these points need extensive elaboration which we cannot give here; all that I want to do for now is to draw out the two corollaries: (a) since a fair rental will be a function of the possible uses of the resource, and since these will normally be greater in an affluent society, an affluent society should, all other things being equal, provide a higher level of payments or benefits than a poor society; (b) since, for any given society at any given time, the amount in the insurance fund is set independently of the number of recipients, the level of payments or benefits will vary in proportion to the number of recipients.

All of this means that we cannot say, in advance, whether the redistribution of wealth called for by the theory of justice will be sufficient to cover the cost of providing all basic needs for all who are indigent. It will depend upon the level of affluence of the society and upon the number of indigents. Suppose then that you were one of the original contractors, trying to decide whether or not you should receive your insurance payments, if you need them, in the form of a cash payment or in the form of in-kind services. This problem will be of special importance to you in the many cases (probably the vast majority of cases in the actual history of mankind) in which the insurance payments will not fund all of your basic needs. It seems to me that you would have a powerful argument for asking for cash payments rather than in-kind services. Choices will have to be made as to which needs are going to be satisfied and at what level. Which choice is best for you is probably different than which choice is best for anyone else. If the insurance payments are in the form of provisions of in-kind services, the choice is not likely to reflect the best mixture for you. If, however, the insurance payments are in the form of cash benefits, you can use that cash to pay for a mixture that seems best for you. So, it would seem that the initial contractors would, as rational agents, agree to the insurance payments as cash payments.

If this argument is correct, then most of our current redistributive programs including Medicare and Medicaid are not properly structured. To begin with, they all suffer from not being part of an integrated redistributive program. We cannot determine the level of funding appropriate for each, for the level of funding is, if our theory is correct, only determinable for the whole of our social insurance program. Not surprisingly, then, some Americans find a particular program too costly, while others find it inadequate.

Secondly, the largest of them (the medical programs and the food programs) provide in-kind services, and there is no reason to suppose that the mix of current services provided is the best for the people in question. For example, many of the recipients would probably be better off with a less elaborate medical program and a better housing program. Finally, many of the eligibility requirements reflect prejudices and misconceptions rather than any notions of justice.

Our society has consistently preferred piecemeal categorial programs to any program of social insurance for all the needy. If my arguments are correct, then the theory of justice would seem to suggest that this preference is mistaken. Actual experience seems to second this suggestion, but that is an additional argument. I conclude for now with the claim that there is only the general problem of the haves and the have-nots, and that we've gotten into trouble by thinking that there are more specialized problems such as that of justice in health care.

Rice University
Houston, Texas

NOTES

[1] Other early versions of this theory are to be found in two forthcoming essays, 'Work Requirements and Welfare Rights' (to appear in a volume edited by the Center for Philosophy and Public Policy), and 'Quasi-Libertarianism and the Laetrile Controversy' (to appear in a volume edited by the Hastings Center).
[2] Earlier versions of this idea are to be found in [3] and [4].

BIBLIOGRAPHY

1. Locke, J.: 1952, *A Second Treatise of Government*, T. P. Peardon (ed.), Bobbs-Merrill, Co., Indianapolis.
2. Nozick, R.: 1974, *Anarchy, State and Utopia*, Basic Books, New York.
3. Ogilvie, W.: 1781, *Essay on the Right of Property in Land*, J. Walter, London.
4. Paine, T.: 1798, *Agrarian Justice*, T. G. Ballard, London.
5. Plattner, M.: 1979, 'The Welfare State vs. the Redistributive State', *The Public Interest* 55, 28–48.

KIM CARNEY

COST CONTAINMENT AND JUSTICE

I. INTRODUCTION

For some years concern about the health sector has been expressed in terms of two major problems confronting the sector: unequal access to health services and rapidly rising costs. More recently, the lion's share of attention has been directed to costs, and the catchword of policy makers, planners, and administrators has become 'cost containment'. It is not that problems relating to access have been solved. To the contrary, despite improvement in the access of the nation's low income and minority families to health services, problems, including numerous inequities, of access persist. Nevertheless, the priority of cost concerns is clear. The purpose of this essay is to explore the meaning of cost containment and its implication for health services delivery with particular reference to the impact on justice within the health care system.

Reasons for Concern With Cost Containment
It has been almost a commonplace for several decades to describe the health sector as being in a state of 'crisis'. An important element of the crisis relates to costs. Rapid increases of prices of existing services, increased utilization of those services, and the introduction of new and often costly, technologically sophisticated services have all contributed to the problem. Per capita spending for health has risen from less than thirty dollars in 1940 to over $ 860 in 1978 (see Table I). More telling than per capita spending are figures expressing spending for health related goods and services as a share of Gross National Product (GNP). This share doubled between 1955 when 4.5 percent of GNP went for health related items and 1978 when the corresponding figure was 9.1 percent of GNP.

There is, of course, no percentage of GNP that is widely accepted as appropriate for health related spending. Spending varies across countries. In 1975 Japan spent about four percent of GNP for health related items while Sweden and the Netherlands spent 7.3 percent of GNP ([5], p. 10). Nevertheless, the concern in the U.S. grows that, if we are not spending too much, we are increasing spending at too rapid a rate. The question of whether we are

TABLE I

Per capita national health expenditures, by source of funds and percent of gross national product, selected years ([3], p. 22)

	Per capita spending			Spending as a percent of GNP
	Total	Private	Public	
1940	$ 29.62	$ 23.61	$ 6.03	4.0
1950	81.86	59.62	22.24	4.5
1955	105.38	78.33	27.05	4.4
1960	146.30	110.20	36.10	5.3
1965	217.42	163.29	54.13	6.2
1970	358.63	227.71	130.93	7.6
1975	604.57	348.61	255.96	8.6
1978	863.01	512.62	350.40	9.1

getting our money's worth for our health dollar is raised ever more frequently. One hears the following: why don't we spend more on preventive care and less on sickness care? Why do we devote so much money to lifesaving technologies only to sustain life in circumstances where it appears that conscious human life can not be restored? Why do we spend ever more money for the training of physicians only to have medically underserved areas persist? And so on. It is troubling that most western European countries, as well as Japan, all of whom spend a lower percent of GNP than the U.S. on health services, have lower infant mortality rates than the U.S.[1]

The growing share of total GNP devoted to health reflects prices in the health sector as well as the quantity, and changed character of health services. The changing prices can be seen by an examination of the Consumer Price Index (CPI). The CPI increased by 81.5 percent between 1967 and 1977 while the index for all medical care rose by 102.4 percent during the same period. Hospital prices, reflected in the indices for semiprivate room and operating room charges, were a major factor in the overall medical care price increases (see Table II).

Other efforts to establish the role of price increases in the health sector produce results that are consistent with the CPI. The developers of a national hospital input price index found that from 1970 through 1978 the average annual increase in input prices was 8.0 percent compared with a 6.6 percent increase in the CPI ([2], p. 37). In a study of costs of treating a group of illnesses, Scitovsky and McCall found that treatment costs rose faster than the medical care component of the CPI ([8], p. 10).

TABLE II

Consumer price index, all items, and selected health related items, 1940–1977 (1967 = 100) ([9], p. 412)

	1950	1955	1960	1965	1970	1975	1977
CPI, all items	72.1	80.2	88.7	94.5	116.3	161.2	181.5
Less medical care	—	—	89.4	94.9	116.1	160.9	180.3
All medical care	53.7	64.8	79.1	89.5	120.6	168.6	202.4
Medical care services	49.2	60.4	74.9	87.3	124.2	179.1	216.7
Hospital service charges[1]	—	—	—	—	—	132.3	164.1
Semiprivate room	30.3	42.3	57.3	75.9	145.4	236.1	299.5
Operating room charges	—	—	—	82.9	142.4	239.4	311.3
X-ray diagnostic series	—	—	—	90.0	110.3	156.2	189.4
Professional services:							
Physican fees	55.2	65.4	77.0	88.3	121.4	169.4	206.0
Dentist fees	63.9	73.0	82.1	92.2	119.4	161.9	185.1
Routine laboratory tests	—	—	—	94.8	111.4	151.4	169.4
Drugs and prescriptions	88.5	94.7	104.5	100.2	103.6	118.8	134.1

[1] January 1972 = 100 (index introduced at that time).

Voices questioning the escalation of health spending are becoming louder and more insistent. Industry, which shares the cost of health insurance, and unions, which negotiate for health benefits, Congressmen, whose voting record on health measures is public information, and consumers, who ultimaterly pay for all health services, are increasingly questioning growing spending for health.

Cost Containment: What Is it?

One way to understand the push toward cost containment is that it reflects a growing awareness of the scarcity of resources, a new understanding of a universal problem. Although discussions of cost containment are appearing with increasing frequency in the literature, the term itself is not often clearly defined. Cost containment, as employed here, involves efforts to stem the growth in spending while maintaining an effective health care system. In this context, efforts to reduce health spending without thought of consequences are excluded. Advocates of cost containment are often persons resisting a simplistic, budget-paring approach; in contrast, they support policy efforts to improve access to services and to improve quality of services.

It is becoming apparent that further pursuit of the positive goals of public health policy, ensuring that adequate and appropriate services are available on reasonable terms to all members of the population, may depend on the success of efforts to gain control over the trend in health care costs ([10], p. 1).

The focus of cost containment policy is upon total expenditures for health related goods and services. Total expenditures are, of course, the sum of the amounts spent for each good and service; that is, the sum of the product of price times quantity for each type of good or service. Algebraically, total expenditure is:

$$TE = \sum_{i=1}^{n} p_i q_i, \quad i = 1, 2, \ldots, n.$$

where TE is total expenditures, p_i is the schedule of prices of each of n different goods and services, and q_i is the schedule of quantities utilized for each of the n different goods and services.

Cost containment may, therefore, be implemented by (1) restraining price increases, or even lowering prices, for individual goods and services, (2) limiting the quantity utilized of some or all goods or services, and (3) changing the mix of goods and services utilized by substituting a less expensive good (service) for a more expensive good (service). Clearly a wide variety of policies fall under the heading of cost containment.

The price of a good or service might be reduced by the adoption of more efficient production techniques. The quantity utilized of a particular good or service might be reduced by some rationing device or by somehow shifting the demand for that good inward. Or, a less expensive good might be substituted for a more expensive one as a result of price incentives. In practice proposals cover the full range discussed here. There is, in terms of the above formula, attention paid to both the p's and q's.

In any sector with a rapidly evolving technology the q_i may change dramatically from one time period to another. This is certainly characteristic of the health sector. Services disappear (the q becomes zero) and new services are added. When the schedule of goods and services changes, it is difficult to assess the meaning of rising prices. That is, a service in 1979 may reflect a considerable quality increase when that service is compared with the most relevant service in 1969. If there has been a change in quality, a higher price may only reflect the additional quality, not a basic increase in price. Any effort to determine what has happened to prices, or to determine whether policies have actually contained costs, in the health sector is faced with these difficulties. Despite the difficulties, it is essential to monitor the success of efforts to contain costs as well as the costs themselves.

Justice in the Health Sector
While it may be possible to define the conceptions of social justice that should be brought to bear in the health care system, it is more difficult to determine how these concepts should be applied, and vastly more difficult to fashion policy that will result in a system consistent with the conceptions. Outka [6] suggests that two among the standard conceptions of social justice are relevant when considering crisis[2] health care. These are:

(1)   To each according to his needs.

(2)   Similar treatment for similar cases.

In trying to relate these two conceptions to the health sector, an immediate response is that great weight falls on the first one of the two because of the immense number of dissimilar cases. Cases vary by age; similar treatment for infants, adolescents, and the elderly is clearly inappropriate. Cases vary by sex; treatment for women during childbearing is necessarily unlike other types of treatment. Cases vary by occupation; professional football players, coal miners, and wheat farmers constitute obviously dissimilar cases. Cases vary by region of the country; the incidence of cancer, for example, varies by region implying dissimilarity. Cases vary by socio-economic level; both health status and the understanding of the requirements for maintaining health appear to vary with socio-economic level — yet another example of the unlikeness of cases.

Clearly great weight, then, in the application of these conceptions to the health sector falls on the definition of needs. Needs differ between different cases as suggested above and also differ within the subsets discussed. A person may go for several years with no need for crisis health care only suddenly to face a severe, lingering illness requiring costly care. It is, of course, the uneven and unpredictable distribution of need for care, coupled with the high cost of that care, that stimulated the development and spread of private health insurance.

The concept of need must be considered at two levels. First, is there a need for health services, that is, is the condition of the patient such that services are warranted? Secondly, what services are needed? There is probably more agreement about the existence of need than there is about appropriate services to meet the need. However, since the patient is frequently the one to initiate the encounter with a physician at the outset of a crisis situation, there remains considerable variety in the severity of need that triggers such an encounter. Once the patient makes contact with the physician there is the significant possibility of disagreement among physicians about the appropriate treatment — about, if you would, the needs of the patient. Treatment is composed of an enormous bundle of services, including diagnostic, remedial, and palliative procedures, which can be assembled in virtually an infinite number of ways. Disagreement among physicians about needed services is compounded by different attitudes among patients toward treatment modes. Thus, in practice whether or not a patient obtains needed services is influenced at several points: the patient's decision to seek care, the physician's diagnosis and recommended treatment, and the patient's response to the physician's recommendation.

While the focus in this discussion is upon health services, it should be kept in mind that the interest in health services is a derived interest — derived from the more basic concern for health status. The underlying concern is with the healthiness of an individual and of society, not the number of physician encounters per capita per year. Presumably, the goal in a just society is to achieve a high and equitable level of health status. Equal health status is, of course, beyond the power of health services to produce. It is, likewise, impossible to achieve equal health status for persons of the same age. An appropriate goal in a just society might be that of achieving an equal probability of good health for persons in different subsets of the population. We might aim for similar infant mortality rates for white, black, native American, Chicano and other babies. Or we might aim at a like life expectancy for persons in such groups. While goals such as these are appropriate for a just society, these goals cannot be met simply by policies related to health services. Health status, including infant mortality and life expectancy, varies with a number of factors: genetic characteristics, socio-economic factors, and 'lifestyle', in addition to the utilization of health services.

It is also the case that health status is difficult to measure. While certain mortality measures such as infant mortality and life expectancy have long been used as indicators of health status, they fail to reflect all changes in status. If treatment improves life expectancy, life expectancy data would indicate this. If treatment does not lengthen life but does enable the patient to live comfortably and earn a living during the course of a disease, that is, if morbidity is lessened, the mortality measures would not reflect the improvement. Thus, analysts assessing the efficacy of policy measures in the health sector even though concerned with status must supplement status measures with utilization data.

Policy measures designed to improve the justness of the system have dealt with the variety of difficulties discussed above by focussing upon access to health services. The discussion, in turn, examines barriers to access — for example, financial and cultural barriers — and changes in access. Efforts to surmount financial barriers include the provision of direct care in community and neighborhood health centers as well as measures to help eligible consumers pay for services purchased in the private market for such services. Medicare and Medicaid assist the consumer to pay for services obtained through private physicians. Outreach programs have been implemented in an attempt to overcome knowledge and cultural barriers. Bilingual programs have been adopted to recruit physicians for underserved areas and to subsidize them for setting up practice in such locations. The rapid expansion of public spending, especially since 1965, reflects various efforts to provide access to health services for low income, elderly, rural, and other populations.

The focus in this study so far as justice is concerned will be, consonant with much of the policy literature, upon access to care. That is, what will be the likely impact of various cost containment proposals on access to care?

II. CHARACTERISTICS OF THE HEALTH SECTOR

Access to Services

On numerous occasions concern about the capacity of the health delivery system to meet the needs of certain groups — low income, minority, and rural families are often mentioned — is expressed. One such period, triggered by the high proportion of young men ineligible for the draft owing to physical disabilities, was immediately after World War II. Another period was in the early sixties following a decade when the infant mortality rate in the U.S. was relatively stable at a time when other developed nations were experiencing sizeable declines; that was also the period of the war on poverty.

There is a considerable body of evidence that access to care has improved especially since 1965. Medicare and Medicaid both date from that time. And almost simultaneously the first neighborhood health centers were established by the Office of Economic Opportunity.

Utilization data today are more nearly equal when comparisons using income, place of residence, and race are employed than they were prior to 1965. For some low income populations, utilization is higher than for other income groups. Health status, as indicated by mortality measures, appears to be more nearly equal than earlier. Reduction in the infant mortality rate for nonwhite infants has been greater than for white infants; life expectancy for nonwhite females has increased substantially more than for white females. The increase for nonwhite males has been slightly less than the rate for white males (see Table III).

TABLE III

Infant mortality and life expectancy at age twenty by race, selected years and percentage change, 1960–79 ([9], pp. 174, 176–177)

Life expectancy at twenty	1960	1970	1979	% Change 1960–79
White				
Male	50.1	50.3	51.6	+ 3.0%
Female	56.2	57.4	58.7	+ 4.4%
All Others				
Male	45.5	44.7	46.8	+ 2.9%
Female	49.9	52.2	54.9	+10.0%
Infant mortality				
White	22.9	17.8	13.3	− 41.9%
All Others	43.2	30.9	23.5	− 45.6%

While there has been improvement in access to services since 1965, the system has considerable distance to go before it can claim to be equitable. Even neighborhood health centers, which have performed admirably in improving access for an eligible population, raise a concern about justice for a similar noneligible population. Since eligibility is usually defined on the basis of residence, persons with similar characteristics residing in and out of the entitlement area are treated differentially.

Programs which assist persons to pay for care purchased in the private sector should, in one way, be able to deal justly with persons. For example, entitlement could be defined in terms of income, but instead of using a single cutoff point for eligibility, a sliding scale of payment based on income is possible. However, neither Medicaid nor Medicare function in this fashion and each program has created, while improving access, new problems of justice. Medicaid raises numerous problems. Not only is there no gradual cutoff for eligibility, but there is a significant difference in the treatment of individuals on a state basis. Medicaid is, in fact, a state program which the federal government helps to finance. Thus persons from like circumstances are treated differently in different states. Generally, the southern states with lower than average per capita income, and thereby a greater need for publicly funded health benefits than states with higher income, provide very limited Medicaid benefits. In contrast, New York and California provide sizeable Medicaid benefits.

Medicare is a uniform national program with similar benefits in all states. Nevertheless, marked disparities in the utilization of health services by Medicare recipients persist because persons from unlike circumstances are treated alike. Low income persons on Medicare — unless eligible for Medicaid also — have the same fees to pay as higher income persons and, therefore, utilize services at a significantly lower rate.

It is the case that the improvement of access to health services since 1965 has been costly. A major reason for the rapid expansion of health costs since that date has been the growth of federal expenditures for health which are designed to improve access. Undoubtedly, the growth in federal expenditures for health is a major factor in the current concern for cost containment. The challenge is to contain costs without worsening the access of groups whose position has improved in the last fifteen years.

Economic Characteristics

A well functioning economic market performs well in the absence of regulation. Competitive forces, Adam Smith's 'invisible hand', bring about an efficient outcome: output produced at the lowest price possible given the costs of production. Supply and demand forces interact to bring about this end.

The market for health services is not a well functioning market; rather it is characterized by a number of peculiarities or market imperfections. The methods of financing health services that have developed over the years interfere with the pricing function of the market. This can be seen on three levels. First a great deal of health care is paid for either by private insurance or the government, that is by third party payors rather than the consumer. This means that the consumer has little incentive to shop around for lower prices, to question whether a particular medication or procedure is actually needed, or to ask whether there is a less expensive alternative. The impact of private insurance is identical with the impact of government programs to finance services for low income families. Since World War II the number of families with private health insurance has grown yearly. Thus, if not now, then, in the foreseeable future, the concern for cost containment would have resulted even if public programs had not been expanded in the sixties.

Secondly, physicians are paid, in the overwhelming number of cases, on a fee-for-service basis. Thus the physician who is a prime factor in determining how much care the consumer obtains has a personal financial incentive to provide repeated care. In private practice there is no institutional pressure to reduce either services or expenditures.

Finally, both physicians and hospitals are reimbursed for services by third parties on the basis of reasonable costs. If either physicians or hospitals can gain acceptance for higher fees, they can increase revenues. So long as reimbursement is on a reasonable cost basis, neither physicians nor hospitals have an incentive to lower costs or seek a less expensive alternative.

Additional characteristics that impede market functioning have been pointed out by economists and others. The consumer or patient is unable to assess the quality of care he is receiving and is strongly dependent upon the supplier of services to tell him when to seek additional services. Further, there are restrictions on entry of providers into the health sector in contrast to the free entry characteristic of the competitive market. Also, many hospitals and other health institutions are nonprofit so that the usual profit motivation for efficiency is absent. In addition, traditional medical ethics prevent physicians and other providers from advertising their charges so that even a price-conscious consumer finds it difficult to take price into account in making his decision.

Intervention in a malfunctioning market is of two sorts: those designed to make the market function better and those designed to regulate the market. To make the market function better implies the removal of market imperfections and the stimulation of competition. Antitrust action is typical of intervention designed to make the market function more effectively. Regulation, usually posited on the belief that the market cannot be improved or cannot be improved in sufficient measure, takes a variety of forms, which may or may not include the direct regulation of prices. Public utilities are regulated industries; in their case price regulation is included.

III. PROPOSALS FOR HEALTH-COST CONTAINMENT

General Characteristics

Cost containment proposals for the health sector reflect the malfunctioning of the market; these proposals fall into the two categories described above — either regulation or stimulation of competition. Regulatory proposals for the sector include the limiting of prices under a form of rate review or prospective pricing and the limiting of the quantity supplied by controlling the number of suppliers or the number of certain types of facilities, such as hospital beds. Attempts to limit quantities supplied by acting upon suppliers are based on the market characteristic discussed earlier, namely that in the health sector the supplier, in particular, the physician, is crucial in influencing the amount of services consumed.

Policies designed to improve market functioning do not attempt directly to control prices but focus on the limiting of quantities through price incentives as well as the indirect effect of competitive forces on prices. These policies may stress the role of the physician in the decision to utilize services or the role of the consumer in the same decision. This decision, whether influenced by provider or consumer determines the quality as well as the type of services utilized. Health maintenance organizations (HMOs) are examples of the attempt to stimulate competition with emphasis on the physician as decision maker. Proposals to broaden health insurance coverage with respect to the variety of services covered while, at the same time, employing proportional copayments are also designed to stimulate competition; these proposals emphasize the role of the consumer as decision maker.

Regulatory Proposals

The bulk of attention of the various regulatory proposals, whether of the sort directly affecting prices or of the sort attempting to limit quantities of services by affecting the supply side, is directed toward hospitals. There are several reasons for this. Hospital expenditures have risen the most rapidly of any health expenditure (see Table II again). These expenditures constitute an important share — 45 percent — of personal health spending. There are, of course, many fewer hospitals than physicians, making it easier, presumably, to regulate hospitals than physicians.

Efforts to regulate hospital prices can occur at a federal level as under the Economic Stabilization Program between 1971 and 1974, or at a state or regional level. A number of states, primarily located in the northeast, have tried a variety of regulatory approaches. These fall into three categories: (1) rates negotiated on an individual hospital basis, (2) rates negotiated on the basis of a hospital's relative performance, and (3) rates determined by means of a maximum rate of increase. The first two emphasize efficiency while the third emphasizes total expenditures.

Costs differ significantly between hospitals. Reasons for the difference include different wage rates, different levels of efficiency, and different case mixes of patients. Teaching hospitals, with their case mix which includes a higher percentage of unusual and severe cases than community hospitals, face higher costs than other hospitals. With these differences existing, it is not surprising that hospitals, if they must have regulation, would prefer rates negotiated on an individual hospital basis. Although Connecticut, Maryland, Rhode Island, New Jersey, and Indiana have employed this approach, it is difficult to evaluate its impact. At most, the impact appears to be minimal.

Blue Cross of Western Pennsylvania has experimented with reimbursing hospitals based on group performance. Initially, nine groups, based on location and teaching status, were defined and average costs of each group determined; later a more complex point system was also made available. No hospital could be reimbursed at a rate more than ten percent above the appropriate average. There were, however, no incentives to encourage the hospital to stay below 110 percent of the average; that is, there were no rewards to the hospital for performing better than average and no penalty for performing worse than average so long as the performance was less than ten percent higher than the mean. Variations of this approach appear in the academic literature, and Senator Talmadge has proposed a bill of this sort.

The Economic Stabilization Program determined a maximum allowable increase in prices — the third type of price regulation. It appears that price increases were slower during the period under this program although prices surged when the restraints were lifted. Rhode Island and New York have used this approach. While there is some evidence of effectiveness, there is also evidence that both hospital admissions and length of stay were increased. At the outset of such a program the most efficient hospitals were penalized. In addition, there is incentive to admit less costly cases in order to maintain total costs in line and to increase costs to the full allowable limit in order to have a higher base for subsequent years. This type of regulation focusses upon total expenditures and thereby the proliferation of tests and other services.

The major tool for restricting the supply of hospital beds is Certificate-of-Need (CON) review. While CON is not limited to hospitals, it has predominantly affected them. Under this type of review hospitals and other facilities are required to prepare a formal request including a justification on the basis of need, prior to embarking on any large capital expenditure. CON review is carried out at the regional and state level in a two-tiered process. The initial review is the responsibility of the Health Systems Agency, one of some 200 regional planning agencies; subsequently, the state review agency acts on the proposal.

The Health Planning and Resources Development Act of 1974 (P.L. 93–641), which established the Health Systems Agencies, requires that every state receiving funds under it have a CON program by 1980. Most states already have a CON program, and in some states CON review antedated P.L. 93–641. CON review limits the entry of suppliers and is, of course, the antithesis of competition. Current guidelines issued at the federal level under P.L. 93–641 call for a limit of four hospital beds per 1000 persons. The national bed to population ratio in 1976 was 4.6 beds per 1000, a considerable oversupply in terms of the guidelines. Excess beds cause concern on two scores: they are themselves costly and they serve as an incentive for providing unneeded care, or overdoctoring.

It is not easy to assess the impact of CON review despite a period of experience with it. It is, of course, possible to examine the CON review records to determine the value of proposals that have been denied, but there are problems with this procedure. First, it is not certain that in the absence of review procedures the denied projects would have come to fruition. Second, there is a possibility that as CON procedures mature, their impact is felt prior to formal review, that the existence of such procedures affects the projects submitted for formal review. This is the view expressed in a recent survey of Health Systems Agencies [1].

The rate of expansion of hospital beds has slowed in recent years. The number of beds grew at an annual rate of 2.1 percent in the decades of the forties and fifties, accelerated to an annual rate of 3.1 percent in the sixties, and dropped to an annual rate of 1.8 percent in the seventies ([9], p. 354). Annual rates of change of beds per capita for the corresponding time periods are 0.6 percent, 1.8 percent, and 1.1 percent. While CON review may have played a role in the lower rates of expansion in the seventies, it may be that there was little demand for additional beds. It is also the case that funds for hospital construction diminished in the seventies.[3]

In a study on the effectiveness of CON review between 1968 and 1972 (prior to P.L. 93–641 in states where CON review existed), Salkever and Bice found:

> ... these controls did not significantly alter total investment by hospitals but did alter its composition. Specifically, the findings indicate that certificate of need programs resulted in lower growth of bed supplies and higher growth of plant assets per bed than would have been observed in the absence of controls. These findings were robust with respect to a variety of specifications of statistical models and analytical approach ([7], p. 75).

Although CON review is the only type of supply limitation which has been implemented in a major way, there are other possibilities. CON could expand to appropriateness review with the power of designating existing agencies as unnecessary. Limiting the supply of physicians is, of course, a possible means to limit health expenditures since they generate sizeable expenditures. By one estimate for 1972, a single physician generates spending of $240,000 a year ([4], p. 1). Recent policy has been designed to increase the number of physicians rather than limit it. This is apparent in the increase in enrollment in medical and osteopathic training. The number of students enrolled in these two areas in 1977–1978 was 33.2 percent larger than the corresponding enrollment in 1971–1972. However, it is possible to infer from the limiting by Congress in the manpower legislation passed in 1976 of entry of foreign medical school graduates that Congress is moving away from a 'more is better' approach for physicians. However, as long as the nation is plagued by distributional problems, with rural areas significantly short of physicians, it seems unlikely that policies to limit the supply of physicians will be adopted.

Utilization review involves a direct attempt to limit the utilization of hospital services by establishing reasonable lengths of hospital stay for various conditions. Professional Standards Review Organizations, established under federal legislation with a goal of improving quality of care, have the responsibility for utilization review. A development tied to utilization review is that of skilled nursing facilities, which are designed to provide less sophisticated and less expensive care than hospitals. When a patient recovers sufficiently to leave the hospital but not enough to return home, a skilled nursing facility can provide interim care.

Improving Market Functioning

Proposals for stimulating competition introduce cost consciousness through financial incentives which may be directed toward either the provider or the consumer. The major proposal focussing on the provider is the furthering of health maintenance organizations, or, as they were called prior to the coining of this felicitous phrase, prepaid group practices. The HMO assumes the role of both insurer and provider with the consumer paying a predetermined fee for which he will receive all needed care. The providers become risk bearers obligated to provide needed services. If costs exceed income, providers suffer; if costs are less than income, they benefit.

A number of sizeable HMOs have developed and survived many years alongside the fee-for-service system. These include: Kaiser-Permanente, the Federal Employees Benefit Program, the Health Insurance Plan of Greater New York, and the Group Health Association. Participation, however, is a small proportion of the total population despite recent legislation designed to extend HMO coverage.

Evidence gathered over the years on the effect of HMOs on costs suggests that HMOs provide quality care at lower than fee-for-service arrangements due to the lower rate of hospitalization for HMO members than for others. However, there may well be a self-selectivity at work, that is, persons who opt to join an HMO are themselves either cost conscious or low utilizers of services. If self-selectivity is a factor, cost savings presently existing would not be maintained if a far wider segment of the population were induced to join HMOs.

In contrast to HMOs, policies to make the consumer more cost conscious have been proposed. Generally these proposals would change the nature of the benefit package covered by insurance and the extent and type of co-payments, but would otherwise leave the fee-for-service system in place. Traditionally private insurance covered the more expensive health services, in particular, hospital care. Since insurance is designed to reduce risk to the owner, it is natural that health insurance would cover services whose need places the consumer at greater risk. However, over the years it became apparent that the existence of such coverage provides an incentive for the utilization of covered services in contrast to lower cost, but not covered, services. The standard but important example is the patient asking his physician to put him in the hospital for various tests so that insurance would pay the costs of the tests.

To overcome this type of situation it is proposed to extend the benefit package so that both less and more expensive services are included, i.e., cover the tests whether done on an outpatient or inpatient basis. This would remove the present perverse incentive to consume the more expensive services. In addition, proportional copayments would be required of the consumer, at least until some defined level of out-of-pocket spending triggered total coverage. If the consumer were to pay, say, 15 percent of costs of testing, he would have an incentive to have the tests done on an outpatient basis.

The effectiveness of such an approach depends on how responsive the consumer is to price incentives, that is, how price elastic is his demand. Although a number of researchers have attempted to estimate price elasticity, it is difficult to obtain valid estimates since the relevant variable, out-of-pocket costs, is largely unknown. It is the case that much of medical care is of a non-emergency sort, and that both out-of-pocket costs and income influence the demand for health services.

At the present time the purchase of health insurance is heavily subsidized by income tax policy. A possible corollary to the described modification in health insurance coverage would be a change in tax policy that would reduce or eliminate the current benefits. Not only are present benefits regressive in their impact, but they encourage consumers to purchase more insurance than they otherwise would, which in turn encourages them to utilize more health services than they otherwise would. A possible compromise would be to eliminate tax benefits except on a model policy of the sort discussed here.

Both the HMO and the consumer cost sharing models would encourage the substitution of less expensive for more expensive services to a greater degree than the regulatory proposals (with the possible exception of the method of limiting hospital price increases to a set percentage). Examples of cost-effective substitutes that might result in certain cases are: (1) outpatient for inpatient surgery, (2) nurse practitioners for physicians, (3) skilled nursing for hospital care, and (4) home or community based services for institutionalization.

Impact on Access

As the previous survey of cost containment proposals suggests, the proposals do not single out any particular group from the population to bear an excessive burden. Rather, the proposals would effect the broad population spectrum. Nevertheless, the proposals raise several concerns. With regard to the regulatory proposals the concern is primarily that less articulate groups might fare less well than the rest of the population. The same concern is, of course, applicable to the present arrangements. Of the three types of price regulation examined, the third, which allowed a predetermined percentage increase in prices, is the most troublesome. There is the possibility that the less articulate patients would find themselves receiving fewer services – fewer tests, for example – than other patients as hospitals tried to limit total spending. Similarly, if a community faces supply constraints as a result of CON review, will the more articulate obtain the scarce services?

Competition generally serves all market participants well. However, a couple of qualifications are in order with respect to proposals to increase competition in the health sector. First, these proposals may make the market more competitive, but they hardly produce a competitive market. As far as HMOs are concerned, it appears that a disadvantaged group can be integrated into and well served by an existing HMO. Nevertheless, California experienced some problems with unscrupulous HMOs, set up to serve a low income population, that were reminiscent of the wildcat banks of the Nineteenth Century.

A significant concern arising from the cost sharing model pertains to the extent of cost sharing required of low income families. Changes in the out-of-pocket costs of obtaining health services appear to have a significant effect on utilization of services for low income families. These copayments would need to vary with income, at least with income below a certain level.

On balance, the current proposals designed to encourage cost containment are not troublesome to those concerned with justice. It would appear that careful monitoring of the impact of these proposals could avoid injustice.

In contrast, the more important cause for concern is the current unwillingness to improve the justness of the present system. The most likely occasion for major improvement in the system appears to be with adoption of some form of national health insurance. Whatever the form of national insurance — and the potential range is broad — it offers the opportunity to improve access to health care. Large numbers of low income families currently ineligible for Medicaid could, and presumably would, be brought into such a plan. Payments for services may either be nonexistent or on a sliding scale based on income. Most national health insurance proposals presume a uniformity of coverage which would move away from the present system which tends towards two tiers of care: one for the poor and one for others.

Since the summer of 1974 when adoption of national health insurance seemed virtually assured, the major obstacle to passage is the concern over the impact on spending of such a step. Given the large number of families currently ineligible for Medicaid and neighborhood health centers, any move toward providing subsidized care for these families will inevitably increase spending. Adding this expenditure to those already growing at a rapid pace is presently unacceptable. Thus the injustices of the present system persist. The impact from failure to move toward national health insurance — the impact of inactivity — is more deleterious than the impact from the move toward cost containment.

University of Texas at Arlington

NOTES

[1] Although there is some objection by American physicians about the comparability of data, it appears there is a clear disparity in these rates with the U.S. coming off second best.

[2] Outka does not dismiss the importance of preventive care but believes the crisis situation shows the 'special significance' our society attaches to health care. The discussion here is also in terms of crisis care since that is the more difficult type of care around which to structure policy; the addition of preventive care to the policy discussion offers little problem.

[3] The Hill-Burton function of financing health facility construction was absorbed in P.L. 93—641, but funds for construction had declined prior to this change.

BIBLIOGRAPHY

1. Anon.: 1970, *Second Report on 1978 Survey of Health Planning Agencies*, American Health Planning Association, Washington, D.C.
2. Freeland, M. S. *et al.*: 1979, 'National Health Care Hospital Input Price Index', *Health Care Financing Review* **1** (Summer), 37–61.
3. Gibson, R. M.: 1979, 'National Health Expenditures, 1978', *Health Care Financing Review* **1** (Summer), 22.
4. Lyle, C. *et al.*: 1974, 'Cost of Medical Care in a Practice of Internal Medicine', *Annals of Internal Medicine* **81** (July), 1–6.
5. Organization for Economic Cooperation and Development: 1977, *Public Expenditure on Health*, OECD, Paris.
6. Outka, G.: 1974, 'Social Justice and Equal Access to Health Care', *Journal of Religious Ethics* **2**, 11–32.
7. Salkever, D. S. and Bice, T. W.: 1979, *Hospital Certificate-of-Need Controls*, American Enterprise Institute for Public Policy Research, Washington, D.C.
8. Scitovsky, A. and McCall, N.: 1975, 'Changes in the Costs of Treatment of Selected Illnesses, 1951–1964–1971', Health Policy Program Discussion Paper, University of California School of Medicine, San Francisco.
9. U.S. Department of Health, Education and Welfare: 1978, *Health: United States, 1978*, National Center for Health Statistics, Publication No. (PHS) 78-1232, December.
10. U.S. Department of Health, Education and Welfare: 1977, *Summary of Grants and Contracts Active on September 30, 1977*, Office of Health Research, Statistics, and Technology, Publication No. (PHS) 70-3234, p. 1.

KAREN LEBACQZ

JUSTICE AND HUMAN RESEARCH

I. INTRODUCTION

Since the Nuremberg Code, two ethical principles have become firmly established as a basis for research using human subjects. 'Respect for persons' requires that the autonomy of the individual be safeguarded — for example, by ensuring 'informed consent' and respecting the subject's desire to withdraw. 'Beneficence' requires that good be done and harm avoided, and is expressed in requirements to exclude certain harms, to do only significant research, and to balance harms and benefits [16].

Until recently, however, discussions of research paid scant attention to the requirements of a third basic principle. 'Justice' requires a fair distribution of burdens and benefits in a social system. It deals with the comparative treatment of persons: who should bear burdens and who should receive benefits [3].

Such decisions are an intrinsic part of the research enterprise. Who shall be solicited as subjects? Will they be remunerated? Should injuries be compensated? These decisions deal with the distribution of burdens and benefits. They raise questions of justice. It is therefore surprising that so little systematic attention has been paid to elaborating a theory of justice for the research setting.

It may have been thought that careful attention to the other principles would obviate the need for a theory of justice. In his seminal essay on research, Hans Jonas [5] derives a system for the solicitation of subjects from the principle of respect for persons. The 'wrong' of treating others as means to the researcher's ends is rectified by the subject's 'identification' with the purposes of the research and will to be involved. A scale of 'descending permissibility' for solicitation of subjects is thus established in accord with their ability to understand the purposes and methods of research. First to be solicited are members of the research community; last are the sick or those who are particularly vulnerable to abuse. It seems, then, that the question of distribution of burdens and benefits of participation might be resolved without explicit appeal to issues of justice.

Recent controversy has brought issues of justice sharply into view, however.

Hepatitis studies on retarded children at Willowbrook School, contraceptive studies on multiparous Hispanic-American women in Texas, syphilis studies on indigent Blacks in Tuskeegee — all have sparked public indignation not simply because of failures in consent or unacceptable risk-benefit ratios. The root issue has been justice: concern about the use and abuse of vulnerable or defenseless persons.

Against this background, the National Commission for the Protection of Human Subjects of Biomedical and Behavioral Research (the Commission) and the HEW Secretary's Task Force on the Compensation of Injured Research Subjects (the Task Force) were established. The Commission was charged with responsibility for determining guidelines for the selection of subjects; the Task Force with responsibility for policy on compensation for injury. Their combined efforts represent an initial attempt to formulate a theory of justice applicable to the research setting.

II. JUSTICE: BASIC REQUIREMENTS

In what does justice consist? What constitutes a fair distribution of benefits and burdens? Formally expressed as 'give to each what is due' or 'treat similar cases similarly', justice requires minimally that persons not be treated arbitrarily. Burdens and benefits must be distributed fairly, in accord with rules about what is 'due' and with attention to relevant similarities and differences [3].

Much depends, then, on what similarities and differences are considered 'relevant' or what claims render something 'due' to one. Need, merit, effort, productivity, utility, and other grounds all have their advocates [19]. In Western democratic tradition, several major criteria emerge as the basis for fair distribution.

First is the fundamental requirement of equality. Barring special circumstances, persons and groups should bear the same burdens and reap equal rewards. Thus, equality is not merely a procedural requirement (equal regard or consideration) but is a substantive one as well. Inequalities in distribution must be justified.

Some such inequalities are generally considered justified — for example, rewards for effort or productivity, or higher payment for goods and services that are scarce [20]. Thus, concerns for *merit* and for *public utility* may justify an unequal distribution.

But what of those who, through no fault of their own, are unable to be productive or to exhibit useful skills? Special need or disadvantage is also considered a 'relevant' criterion that entitles one to a greater share of the benefits and a lesser share of the burdens.

Finally, justice requires punishment for wrongdoing and compensation for injury [1]. Those who have wronged another are thought to be 'due' some form of penalty or burden, while those who have been harmed by another are considered to be owed some compensation. Whether such questions are distinguished as 'retributive' or 'compensatory' or whether they are simply lumped under the rubric of 'distributive justice', they involve the distribution of burdens and benefits.

In general, therefore, justice requires an equal distribution of benefits and burdens except where unequal distribution is justified by considerations of merit, social utility, need, or prior harm. Justice refers not simply to the distribution of goods among individuals but also to social patterns and practices [18]. Fair treatment of individual persons is one aspect of justice; the way in which social institutions distribute goods over time to persons and groups is another. There are both 'individual' and 'social' aspects to justice [16].

III. APPLICATIONS: SELECTION OF SUBJECTS

Some applications of these general principles to the problem of selection are immediately obvious. However, the Commission addressed a number of special problems in the interpretation of principles for certain subject groups. A quick review of the Commission's work will indicate general applications and specific interpretations.

Equal Treatment: The Fetus as a Test Case

No person or group should bear disproportionate burdens of participation in research; such burdens should be broadly distributed among racial, ethnic, sexual, and socio-economic groups. 'Equal treatment' implies equal use of population groups as subjects and equal access by groups to the medical advances deriving from research. Hence, at all points the Commission required equitable selection of subjects.

This principle was tested in the Commission's deliberations on the use of the fetus [12]. Is it permissible to use fetuses scheduled for abortion (FA) in research that would not be permitted on fetuses going to term (FT)? To do so seems at first glance to treat those who are scheduled to die as less than 'equal', to deprive them of equal protection. If so, then it is clearly a violation of justice.

However, much depends on the meaning of 'equal treatment'. If it means subjecting all those in the same class (i.e., all fetuses) to the same *procedures*, then one should not use a procedure on an FA that one would not use on an FT. But if it means exposure to the same degree of *risk*, then if there is reason to believe that the risk is different for FAs and FTs from the same procedure, they might be subjected to different procedures without violating the requirements of justice.

Suppose for example that two adults are solicited for research on smog levels. Both will sit in a dusty room and various breathing measurements will be taken. On the surface, it seems that both are treated 'equally' so long as the same procedures are used. But suppose one suffers from asthma while the other does not. In this case, the *risk* to the two from the same *procedure* is not equal. 'Equal protection' here requires protection from risk, not simply use of the same procedures.

Analogously, FTs could be at considerably more risk than FAs from the same procedures. Suppose research is being done to discover whether a drug crosses the placenta and harms fetal tissue. The fetus aborted within a few weeks of injection of the drug may show relatively little damage from the research procedure, whereas the fetus brought to term might suffer life-long disability. The risk to the two is not the same from the same research procedure.

Of course, there is a slight possibility — estimated at less than 1% — that the woman will change her mind, refuse the abortion, and that the FA will become a damaged newborn. Still, its chances of this harm are about 1% of the FT. Hence, the risk to the two groups differs.

Thus, the Commission recommended that both FAs and FTs be subjected to no more than 'minimal' risks; but it anticipated that this might permit some research to be done on FAs that would not be done on FTs since the risk to the two groups from the same procedure may differ. The Commission thus established that 'equal treatment' means equal protection from risk and that the fact that one is dying or condemned to death does not remove the requirement of equal treatment.

Merit: Social Justice and 'Therapeutic' Research
As noted above, the requirement of equal treatment can be modified by considerations of merit: those who have borne burdens should reap benefits. Two caveats must be entered regarding the interpretation of this modification.

First, it is probably impossible to require that on an *individual* basis those who bear burdens of participation in research should be the first to reap benefits of new medical advances. Some who participate in research will not live to see the results. Many important breakthroughs in research derive from 'basic' research or from unforeseen extensions and applications. Hence, it is unrealistic to require that specific subjects always reap the first benefits derived from research. However, justice may be applied in its *social* aspect here: the class of research subjects in general should be among the first to receive benefits of new therapies. In this regard, it must be noted that much research currently is performed in teaching hospitals using primarily poor or indigent subjects. If benefits of new therapy go first to the rich or to private patients, then such a system is patently unjust, for the class which bears the burdens does not reap the benefits.

Second, one must beware of supposing that therapy given during the course of research constitutes sufficient 'benefit' to offset the burden of participation in research. It is tempting — but fallacious — to think that the requirements of justice will be met so long as those who participate in research receive some form of therapy (commonly called 'therapeutic' research). The Commission rejected the designations 'therapeutic' and 'nontherapeutic' research and warned against 'package analysis' that lumps the risks from both research interventions and therapeutic interventions and then balances those risks against similarly lumped benefits [14]. It required that risks from research interventions be no more than minimal except under carefully guarded conditions. This reflects in part the Commission's conviction that the benefits from research derive primarily from the research enterprise as a whole and that therapeutic maneuvers offered during the course of research must not themselves be considered a 'benefit' of research but must be considered separately [11].

The notion of 'merit' in justice thus has two primary implications for the selection of subjects: it requires attention to the pattern of distribution and the use of subjects from classes who do not reap benefits of medical advances, and it runs the danger of leading IRBs to assume that research protocols that include therapeutic interventions for the subjects will automatically meet requirements of justice.

Disadvantaged: Children and 'IMIs'

Attention to special need or disadvantage also suggests some modifications in the distribution of burdens and benefits. Those who are disadvantaged should bear burdens of participation in research only where more advantaged populations have already participated or are not appropriate as subjects because of the specific design of the research. For example, children should be involved only after adults (unless the disorder or process to be studied is one that occurs only in children) [14]; similarly, non-institutionalized persons should be used before those confined in institutions [15]. Something akin to Jonas's descending order of permissibility for involvement of subjects is therefore required by the claims of justice deriving from special need or disadvantage. Where such subjects must be used, additional safeguards and protections such as consent monitors or court review may be required.

Indeed, one advocate has argued that children and the mentally infirm should never be solicited for research that is not designed to foster their own well-being. Ramsey [17] contends that because such subjects cannot give consent, to use them in 'non-therapeutic' research is to treat them simply as means to another's ends and hence is to violate the principle of respect for persons.

There are numerous problems with this argument. First, as noted above, the designations 'non-therapeutic' and 'therapeutic' research are misleading, for even in 'therapeutic' research (in which research is combined with the provision of therapy) *some* procedures are done that are not for the good of the subject alone. Other objections include the argument that the child is not yet fully a 'person' and thus that the principle of respect for persons does not apply or the argument that a person's proper ends include more than simply health and well-being and thus that 'non-therapeutic' research may not violate the unconsenting subject's proper ends [6]. As an argument based strictly on respect for persons, Ramsey's contentions falter.

However, a similar argument might be grounded in concerns for justice. Since justice requires that vulnerable subjects should bear fewer burdens and reap more benefits, one might find here sufficient grounds for prohibiting the use of disadvantaged subjects in research presenting specifiable risks and offering the subjects no benefits. Indeed, in general the Commission recommends that no more than minimal risk should be presented to disadvantaged subjects from research interventions ([14, 15]).

But the Commission also permitted 'minor' increases in risk provided the experiences were commensurate with those the subject faces in the normal course of treatment [14]. Limits on both the amount and the type of risk are thus imposed, but important research involving no therapeutic interventions is not altogether foreclosed.

At stake here is a delicate balancing of the demands of social and individual justice. Justice might be violated on an individual level if vulnerable persons are permitted to bear burdens without reaping benefits; however, the *class* of subjects does reap benefits from the research. Moreover, these benefits may be obtainable for the class only if some of its members are put to risk. Only in cases where the research offers promise of contributing significant knowledge leading to the amelioration of, e.g., childhood disease, would it be permissible to use disadvantaged subjects. Here individual subjects would be permitted to bear more burdens and reap fewer benefits provided the benefits to the class are substantial. Even so, limits are set on the amount and type of risk to be borne by the individual. Thus, concerns of social and individual justice are balanced.

A final feature in the application of justice to the case of those who are disadvantaged emerged in the Commission's deliberations on both children and those institutionalized as mentally infirm. The very labelling of a class raises questions of justice. First, in the case of both children and 'the mentally infirm' it must be recognized that the label alone does not portray the range of capacities and incapacities within the class. While many commentators appear to picture the prototypical child at around the age of five, in fact a fifteen-year-old is also legally a 'child'. Hence, the label points to some common aspects of disadvantage that are shared (e.g., lack of certain legal rights) but obscures important distinctions, within the class. If 'disadvantage' is a morally relevant criterion, then it should function to make important distinctions not only between classes but also within each class.

Second, the more troubling is the possibility of inaccurate application of a label. The label 'mentally infirm' in particular brings to bear an interpretive framework that carries burdens for the individual so labelled. In recognition of the possibility of error and in order to avoid prejudging the mental capacities of those assigned to mental institutions, the Commission adopted the phrase 'those institutionalized *as* mentally infirm' [15]. The mere fact of institutionalization serves to render such persons 'disadvantaged'; justice therefore requires that they not bear undue burdens, including those associated with labelling.

Considerations of need or disadvantage thus modify the strict requirements of equality in a number of ways: a 'descending scale' of involvement is established from the more advantaged to the less advantaged, special protections may be required, limits on both amount and type of risk may be set, and attention must be paid to small ways in which additional burdens might be heaped on disadvantaged subjects.

Equality and Disadvantage in Conflict: Prisoners
The different requirements of justice occasionally conflict. Such conflict is illustrated in the Commission's deliberations on the use of prisoners in research [13].

On the one hand, prisoners generally live under severe restraint in 'total' institutions isolated from public scrutiny where even daily necessities may be arbitrarily withheld. To this extent, they are a disadvantaged and vulnerable population and justice requires that they be protected from risk. Following such an argument, some commentators argue against the solicitation of prisoners for research.

On the other hand, prisoners are generally competent adults capable of making decisions on the basis of information received. Excluding them from participation in research may constitute a deprivation of freedom and thus violate their 'equal treatment' with other competent adults. Hence, some argue that ensuring prisoners' 'informed consent' to participate provides sufficient protection.

Does justice require that prisoners be protected from risk or provided equal opportunity to participate in research? The first problem here is determining the proper interpretation of the status of prisoners: does the fact of confinement and the burdens it imposes render them a disadvantaged population who should receive extra protection, or does the competence and equality of prisoners with those on the outside render them an 'equal' population who should be able to choose to bear burdens in order to reap benefits?

Complicating this issue is some disagreement about the 'risky' nature of research. Many commentators appear to assume that most research is risky and therefore imposes burdens on vulnerable populations. However, evidence suggests that participation in research may not be as risky as being an office secretary [8]. Prisoners themselves tend to stress the benefits of participation: relief from boredom, an opportunity to make money, and a chance to contribute to society. If participation in the types of research generally conducted in prison provides benefits and does not impose burdens, then prisoners need not be excluded from solicitation even if they are disadvantaged.

The dilemma is further exacerbated by racial imbalances in prison. Prison populations tend to have a greater proportion of those from racial and ethnic minorities and lower socio-economic strata than is found in society at large. Thus, permitting research in prison potentially adds risks to groups that are disadvantaged and raises questions of social justice.

The question of remuneration illustrates the interplay of such questions. Pay for most work in prison is very low. If participation in research is remunerated at this low rate, researchers would appear to be taking advantage of a captive population and violating the requirements of equal treatment vis-à-vis those on the outside. But to remunerate at a higher rate violates equal treatment vis-à-vis others on the inside and may therefore be an unfair inducement to participate. Given the low level of most prison pay, is it possible to design a fair system of remuneration for participation in research?

All of these factors combine to make the question whether prisoners may fairly be selected as research subjects a very complicated one. After months of deliberation, the Commission arrived at the following balance in the requirements of justice.

First, the Commission viewed prisoners generally as competent adults capable of making decisions and hence 'equal' with other adults. However, the Commission was also convinced that the prison environment can undermine freedom of choice. Where options are few, boredom is rampant, pay is minimal, living standards abysmal and self-worth elusive, any activity that offers relief from boredom, adequate pay, decent food and a sense of contribution to society becomes a powerful inducement. Such inducement is not 'coercion' *per se* but can undermine an accurate assessment of risk. To this extent, the prison setting — at least under current conditions — renders prisoners a vulnerable population. The Commission therefore decided that while there is no reason *a priori* to exclude prisoners absolutely from all participation in research, the requirements of justice that apply to vulnerable populations should apply here.

The question of justice among racial and socio-economic groups remains ambiguous. Although there are disproportionate numbers of non-whites and of the poor in prison, most participants in research there are white and relatively well situated. Thus, it does not seem that the poor and non-white groups bear a disproportionate share of the risks of research in prison. However, this does not settle the question of justice. If participation in research is a *benefit* and not a burden, then injustice may exist in the more ready access to this benefit by whites and those better situated. Ensuring equitable selection of subjects within the prison population takes on additional importance in view of this uncertainty.

It is also possible, of course, that non-whites are involved in the more risky research protocols and thus do bear more risks. Allegations that some research done in prison is very dangerous are difficult to substantiate or to refute because of the closed nature of the prison setting and its isolation from the community. The Commission therefore stressed the need for public scrutiny of conditions within the prison setting and called for public policy to stress protection from risks in the meantime. In short, the Commission made its recommendations on the basis of presumptive evidence that prisoners are vulnerable subjects and hence that those norms derived from the principle of justice that apply to vulnerable subjects should apply here.

Thus, less vulnerable subjects should be used wherever possible. Research that must be conducted in prison — e.g., on the impact of the prison environment — is permissible. But other research should be done elsewhere, or strong justification must be given for using this vulnerable population. Reasons of administrative convenience do not suffice.

Research that includes therapeutic interventions for the individual prisoners is also permissible. However, the Commission warned against possible abuses of this recommendation: not everything argued to be 'therapeutic' for prisoners is genuinely therapeutic and justifiable.

All other research may be done in prison only where the coercive aspects of the environment are reduced and controlled. Minimal living standards reduce the incentive to participate in research in order to obtain bare necessities. Public scrutiny and adequate grievance procedures reduce possible abuses of prisoners' basic rights. Such measures were proposed by the Commission in order to reduce the possibility of 'unfair inducement' to participate in research.

The Commission did not make any final recommendations about remuneration, but noted several systems that might avoid the twin dangers of taking advantage of a captive population or creating unfair circumstances within the prison. For example, prisoners could be paid the equivalent of free-living persons, but receive only a portion of that amount directly for themselves and their families; the remainder would go into a fund administered by prisoners for the improvement of living and working conditions in the prison generally. Thus over time a more fair distribution of all goods in prison would result.

Summary

Not all of the Commission's recommendations have yet become regulation, and some have been hotly debated. Nevertheless, the basic interpretations of the requirements of justice established by the Commission for the selection of subjects provide the foundation for such debate: justice requires equal distribution of burdens and benefits and that those who have borne burdens should receive rewards. Considerations of individual and social justice will have to be balanced on occasion, as will the requirements of equal treatment and protection of the vulnerable.

IV. APPLICATIONS: COMPENSATION FOR INJURY

Considerations of reward and reparation raise questions of compensation for injury sustained during participation in research. Minimally, where a specific *wrong* is done to subjects that results in their harm, justice would seem to require restitution of the wrong or compensation for injury [1]. The law makes provision for such occurrences.

But what about cases where there is no specific wrongdoing, but subjects are nonetheless injured during their participation in research. Should such injuries be compensable? It is this difficult question that was put before the Task Force. In particular, the Task Force was faced with two thorny questions: does the fact of having given consent obviate the responsibility to provide compensation, and is compensation required even where the subjects receive therapeutic interventions in the course of research?

It seems at first glance that one's consent to participate in research might obviate any claims that could otherwise be brought for injury. For example, if I choose voluntarily to go on a mountain-climbing expedition, knowing full well the risks, then surely I have forfeited my right to bring a claim for compensation in any case except where there is negligence on the part of the expedition leaders or other participants. Is this not analogous to the situation regarding research?

The Task Force rejected such a line of reasoning on grounds that "injured parties are persons involved in activities sponsored by and directed toward the benefit of society as a whole" ([4], pp. VI-3). In short, subjects take on risks in endeavors supported by the community in order to contribute to the general well-being. Society therefore has an obligation to protect the well-being of the subjects, at least where the research is conducted, supported, or regulated by the federal government. As with those who volunteer for military service, the fact that they have volunteered does not waive their right to seek compensation where they have participated in risky endeavors for the good of the whole.

But then it might seem that subjects would not be owed compensation for their participation in 'therapeutic' research, in which they will receive some form of therapeutic intervention, for here they participate not simply for the good of the whole but out of their own interests as well. Although subjects may benefit from the therapeutic interventions included in a research protocol, the Task Force concluded that they are nonetheless entitled to compensation for *research-related* injury. The key is to distinguish injuries that are research-related, since subjects may be harmed by their diseases and by any standard therapies used to ameliorate such diseases. Thus, it is not only the research that may bring injury in such cases.

The Task Force concluded that justice requires that subjects be compensated for harm which (1) is 'proximately caused' by the research, and (2) exceeds that 'reasonably associated' with the patient's illness and with standard interventions for that illness ([4], pp. VI–9). For example, if loss of hair, vomiting and diarrhea are commonly associated with interventions for cancer, then they are not compensable if they result from an innovative approach to the treatment of cancer. However, if the innovative approach results in disability not normally associated with treatment (e.g., kidney failure), then compensation is owed.

Summary

The Task Force thus achieved a delicate balance between the claims of justice and the assumption of risk taken for one's own benefit by holding firm to the notion of compensation for those risks taken essentially on behalf of others.

Like the Commission's recommendations, however, implementation of these recommendations is not without problems. In response to the Task Force report, DHEW published 'interim final regulations' on November 3, 1978 requiring that, effective January 1979, negotiations for 'informed consent' must include an explanation as to whether compensation and medical treatment are available if physical injuries occur as a result of subjects' participation in research. Several universities have responded by setting up mechanisms to compensate for research-related injury [9]. However, the 'regulations' have also sparked some controversy, and it is not clear what the final outcome will be ([2, 10]).

V. CONCLUSIONS: JUSTICE AND HUMAN RESEARCH

Although both the Commission's recommendations and the Task Force Report have become the subject of some controversy, one thing seems clear: from this point forward, IRBs and investigators will have to pay attention to issues of justice in the selection of subjects and in their treatment during the course of research and afterward. It will no longer suffice to assume that 'informed consent' or attentions to 'risk-benefit calculi' ensure the ethical acceptability of research. The claims of justice must also be met: individuals and groups must be treated equally except where unequal treatment is justified by considerations of merit, need, or prior injury.

Pacific School of Religion
Berkeley, California

BIBLIOGRAPHY

1. Ezorsky, G.: 1972, *Philosophical Perspectives on Punishment*, State University of N.Y. Press, Albany, N.Y.
2. Fishbein, E. A.: 1979, 'Objections to the 'Interim Final Regulation'', *IRB* 1 (March), 9–11.
3. Frankena, W. K.: 1973, *Ethics*, 2nd edition, Prentice-Hall, Englewood Cliffs, N.J.
4. HEW Secretary's Task Force on the Compensation of Injured Research Subjects: 1977, *Report*, DHEW No. (OS) 77–003, Washington, D.C.
5. Jonas, H.: 1969, 'Philosophical Reflections on Experimenting with Human Subjects', in P. A. Freund (ed.), *Experimentation With Human Subjects*, George Braziller, N.Y., pp. 1–31.
6. Lebacqz, K.: 1978, 'Pediatric Drug Investigation: Current Ethical Guidelines', in B. L. Mirkin (ed.), *Clinical Pharmacology and Therapeutics: A Pediatric Perspective*, Year Book Medical Publishers, Inc., Chicago, pp. 279–297.
7. Lebacqz, K.: 1979, 'Fetal Research: A Commissioner's Reflection', *IRB* 1 (June/July), 7–8.
8. Lebacqz, K. and Levine, R. J.: 1977, 'Respect for Persons and Informed Consent to Participate in Research', *Clinical Research* 25 (April), 101–107.
9. Levine, R. J.: 1979, 'Advice on Compensation: One IRB's Response to DHEW's "Interim Final Regulation"', *IRB* 1 (March), 5.
10. Levine, R. J.: 1979, 'Advice on Compensation: More Responses to DHEW's "Interim Final Regulation"', *IRB* 1 (April), 5–7.
11. Levine, R. J.: 1979, 'Clarifying the Concepts of the National Commission', *The Hastings Center Report* 9 (June), 21–26.
12. National Commission for the Protection of Human Subjects: 1975, *Report and Recommendations: Research on the Fetus*, DHEW No. (OS) 76–127, Washington, D.C.
13. National Commission for the Protection of Human Subjects: 1976, *Report and Recommendations: Research Involving Prisoners*, DHEW No. (OS) 76–131, Washington, D.C.
14. National Commission for the Protection of Human Subjects: 1977, *Report and Recommendations: Research Involving Children*, DHEW No. (OS) 77–0004, Washington, D.C.
15. National Commission for the Protection of Human Subjects: 1978, *Report and Recommendations: Research Involving Those Institutionalized as Mentally Infirm*, DHEW No. (OS) 78–0006, Washington, D.C.
16. National Commission for the Protection of Human Subjects: 1978, *The Belmont Report: Ethical Principles and Guidelines for the Protection of Human Subjects of Research*, DHEW No. (OS) 78–0012, Washington, D.C.
17. Ramsey, P.: 1970, *The Patient as Person*, Yale University Press, New Haven.
18. Rawls, J.: 1971, *A Theory of Justice*, Harvard University Press, Cambridge.
19. Rescher, N.: 1966, *Distributive Justice*, The Bobbs-Merrill Company, New York.
20. Vlastos, G.: 1962, 'Justice and Equality', in R. B. Brandt (ed.), *Social Justice*, Prentice-Hall, Inc., Englewood Cliffs, N.J., pp. 31–72.

RONALD M. GREEN

JUSTICE AND THE CLAIMS OF FUTURE GENERATIONS

Must we be morally concerned with the welfare and the health of future generations? Are we required to leave our distant descendants the conditions for a healthy life? And how do our answers to questions like these affect our health care responsibilities to our contemporaries? While a focus on our obligations to the future is a hallmark of much recent ethical discussion — the issues of population growth, energy and environmental protection come immediately to mind — the question of our health-related responsibilities to the future has rarely been explored. Most ethical discussion of health care and justice, for example, has focused on the equitable distribution of present health care resources, especially the right of less-advantaged individuals to medical services.

Important as this question of the right to health care is, a concern with only presently living individuals reflects a kind of chronological parochialness in our thinking about health care as well as an insensitivity to the long term consequences of our policy choices. In the remarks that follow I want to try to break out of this present-oriented context to look at our health care obligations in a more intergenerationally responsible way. To do this we must first traverse some philosophically puzzling terrain. But the effort is ultimately worthwhile, I think, precisely because it can help us with some of the new choices about justice and health care we are being called upon to make today.

I. PHILOSOPHICAL EXPLORATIONS

Behind the question of our health-related responsibilities to the future lies the more basic question of whether we have moral responsibilities to future persons at all. The suggestion that there is a question here may seem odd. Ordinarily we feel somewhat responsible to those who will follow us. But if we seek a clearer understanding of why this is so, especially where distant generations are concerned, our certainty weakens and perplexities crop up that have occasioned considerable philosophical debate ([8, 9, 11–13, 19, 22, 23, 29–31, 33, 37, 38]).

For example, if we believe that moral obligation derives from express commitments we have made to other people or from ties of real sentiment or affection that exist between us, then neither of these seems to furnish a basis for affirming obligations to the more distant future. Since the generations we are talking about are yet unborn, we cannot have incurred obligations to them by word or deed, nor can we think of ourselves as tied to them by natural affection. Such an account of obligation may explain our moral responsibilities to our own children or our children's children, but it breaks down where the distant future is concerned [33]. Not surprisingly, those philosophers who have derived moral obligations from express commitments or real ties, when asked the question 'What do we owe the distant future?' have answered, 'Nothing' ([12, 13]).

Finding this conclusion intuitively unacceptable, other philosophers have pressed more deeply to an understanding of the nature of moral obligation generally. They have been attracted to the position known as utilitarianism which holds that the fundamental aim of morality is to increase human well-being (or to decrease suffering). This is the basis of the well-known utilitarian principle requiring us to promote the 'greatest happiness for the greatest number of people' [28]. In this context the attractive thing about this rule is that it seems to carry within it an answer to our question about obligations to the future, for on the utilitarian account, where or when persons live has no bearing on our obligations to them. If our present behavior causes some distant future persons to suffer avoidable harm, then we are acting wrongly and should alter our conduct. Of course, utilitarians are aware of how difficult such long term calculations can be, but they believe that at least their basic moral position is clear and provides an account of obligation congruent with our intuitive sense of responsibility to the distant future.

Critics of utilitarianism, however, have pointed out some bizarre implications of this position. For example, if our duty is to promote the greatest happiness for the greatest number, why should we not produce more *persons* to experience happiness? A utilitarian policy seems to counsel greatly expanded populations in the future, possibly at some significant loss of *per capita* or average well-being, so long as the sum total of human happiness increases ([36], pp. 414 f.; [42]). Even if this problem could be overcome, as some have tried to do [30], still another difficulty appears when we consider that the moral community identified by utilitarianism is heavily weighted towards the future. We and our immediate descendants compose, at best, several generations, but out beyond us extend countless generations of persons whose welfare must be read into the moral equation. Now, if we are morally required to produce 'the greatest happiness for the greatest number of persons' then *our* needs must pale beside the interests of the future and the most onerous sacrifices on our part would seem to be required. And this seems to be true even if we heavily discount the future because of our uncertainty about it ([13], p. 508). In other words, it seems that to the question 'How much do we owe the future?' the utilitarian must answer, 'Everything'.

I might point out here that this conclusion is not limited to the confines of the academic ivory tower but has been put into practice in certain societies where forms of utilitarianism have been the operative public philosophy. The willingness of nineteenth-century laissez-faire capitalism and some twentieth century Marxist societies to sacrifice the health and well-being of entire present generations to the future shows that the utilitarian position has had practical effect ([3], pp. 273 f.). If we abhor the mentality that lay behind the nineteenth-century sweatshop or that more recently produced the agony of Cambodia, however, we must probably also reject the utilitarian account of our obligations to the future.

The obvious difficulties with either of these extreme efforts to explain our moral relationship to distant future generations has prompted some fresh thinking about basic premises. In different ways, a number of philosophers have come to reject the idea that moral obligation derives either from personal relationships or from a concern with promoting human welfare as a whole. Instead, these philosophers maintain that morality is best understood as a uniquely rational way of adjudicating social conflict. It represents the effort to replace power and force with reason in the settlement of social disputes and it seeks to make productive cooperation possible where conflict would otherwise reign ([12, 43]).

This basic understanding yields in turn a model of moral reasoning in which, to resolve a potential conflict, we are asked to regard it from the point of view of all the contending parties. The aim is to achieve a measure of objectivity and impartiality in judgment. Perhaps the foremost example of this approach is John Rawls's social contract theory of justice [32]. Rawls's theory requires us to regard difficult social and moral decisions as though we are rational voters in a hypothetical contract situation. Within this situation we are encouraged to pursue our interests as best we can but to prevent the distortion resulting from mere self-interest, we are asked to think of ourselves as ignorant of just *who* we are, or, more exactly, of how we alone are affected by the outcome of our decisions. We must ask 'How would I feel about the proposed policy if I could be any one of the many persons affected by my decision? All things considered, what would I advocate doing?' By approaching matters this way, Rawls maintains, we can conceptually overcome the barrier of self-interest and, at least in theory, arrive at principles and policies able to elicit the consent of all.

While Rawls's approach has many complexities, the basic idea is as familiar as the everyday counsel to 'put yourself in the other person's shoes'. In a very general way, moreover, this approach also helps us understand the nature of our obligations to the future without incurring some of the problems associated with the views we have just considered. For example, if we regard morality as a way of rationally settling social disputes, it becomes clear that moral duties are not confined to those persons to whom we have made commitments nor to those with whom we are personally concerned. Potentially, moral relations and duties exist between any persons whose interests might conflict and in any situation where reasoned cooperation rather than rule by force is generally desirable. Certainly these features are present in the relationship between generations. Who would want our dealings across time to be reduced to a Hobbesian 'war of all with all' in which each generation is permitted selfishly to pursue its own interests leaving those who follow to struggle on amidst the debris?

We have obligations to the future, therefore, because it is only proper that a moral relationship exist among those separated in time. But understanding the basic reason for these obligations also sets a limit on them and argues against the excesses of the utilitarian view. For example, the fact that obligations to the future derive not from an abstract concern for promoting happiness but from a desire to prevent conflict between generations means that we are under no obligation to multiply the number of persons alive or to support growing populations. Quite the contrary, since our concern is to act in a way that might be accepted by those real persons who will follow us, for their sake we should probably strive to keep human numbers within a reasonable and comfortable limit ([17], pp. 194–196). Similarly, the root idea of morality as involving the impartial arbitration of dispute in a contract-like situation means that it is probably wrong to require any present generation to sacrifice everything for the future. In keeping with what has been said, each of us is required to ask the question: "Which policy or course of action would I be prepared to advocate if I did not know the generation to which I belonged — if I could be alive now or in any one of the generations to come?" ([15], p. 261). While the sheer bulk of the future makes it reasonable for us to protect our possible interests there, we would certainly not do this to our serious detriment as presently living persons. What impartial rational person, after all, would be willing to pillory himself as a member of some present generation to marginally enhance his possible position in the future?

The outcome of this reasoning is that within a proper understanding of intergenerational morality, not only do we owe something to the distant future, but the distant future owes something to us. At first sight, this seems impossible. How can those living years from now owe anything to generations long deceased? How, given the inescapable directionality of 'time's arrow' ([23], p. 209), can our obligations reach back into the past? Nevertheless, a conception of morality which sees all generations as gathered together in a purely metaphorical contract situation to forge an enduring set of principles for the sake of cooperation makes conceiving such obligations possible. In this timeless place we retreat to in our minds and consciences, we can reasonably speak of duties among individuals who can never know one another in person.

II. THE EXTENT OF OUR OBLIGATIONS

Our obligations to the future, therefore, lie somewhere between the extremes of all or nothing. We need not sacrifice everything for the future, as some utilitarian thought would suggest. Neither may we totally neglect the welfare of distant future generations. But just where on the spectrum between these extremes does our responsibility lie? How much do we owe the future? Initially, one is drawn to the idea that all generations merit equal treatment so that it is our obligation to hand on to our descendants opportunities for well-being substantially the same as those we have received from the past. Certainly, equality is a natural solution to distributive problems where roughly similar parties contend over scarce resources and opportunities. It is also reasonable to suppose that it would be unfair to hand on to the future a diminished net stock of goods and opportunities with human history thus assuming the character of a downhill slide.

In fact, I suspect that equality of condition is too modest a goal to aim at for our descendents. The very imprisonment in time which accounts for the odd quality of cross-generational obligations also lends a peculiar quality to cross-generational cooperation. As is true of all human cooperation, modest temporary sacrifices can yield greater benefit for almost everyone, although in this case the flow of time excludes any present generation from enjoying the fruits of its own cooperative acts. Despite this loss, however, policies aimed at progressively improving the condition of our descendants make good sense from an impartial standpoint. By not limiting ourselves merely to replacing what we have received, by striving to leave our descendents 'better off' than ourselves, we can do our part in nourishing a constantly improving human future. And if every subsequent generation does the same, distant descendants can experience levels of well-being far beyond our own.

These considerations suggest a rough rule of thumb or very general principle to guide our inter-generational moral thinking. According to this principle *we are required to strive to ensure that our descendants are left with the means to a progressively better quality of life than ourselves and, at a very minimum, are not rendered worse off by our actions* ([15], pp. 261 f.). As stated, this principle, raises many questions. For example, what is meant by 'better' or 'worse off' in this context? Are we talking only about the measurable aspects of a standard of living: disposable income, mortality rates or longevity? Or must we also consider less easily defined values such as freedom, opportunity or 'health' in the broadest sense of that term (the World Health Organization definition, for example)? And what are we to say about very complex non-material values such as wisdom or self-contentment? Do these deserve a place in the improved quality of life we must leave to our descendants?

Despite the attractiveness of quantifying modes of analysis, I suspect that a reliable assessment of quality of life must take all these values into consideration. Surely recent experience has taught us that higher levels of disposable income accompanied by environmental degradation and mental stress are not necessarily desirable. In the health care field alone we have recently become aware of the shortcomings of efforts to extend the duration of life with no corresponding attention to the quality of the additional years gained. A sensitive concern for the future, therefore, must consider the whole array of values commonly cherished by human beings. Case by case, future state by future state, we must ask whether the legacy handed on represents an improvement or worsening in the human condition. Possibly the only general guide moral theory can lend to this process of evaluation is the emphasis on self-interested, knowledgeable but *impartial* reasoning. Specifically, we must ask: 'If I were to live in some future generation, and if I could be any representative individual in that generation, would I have reason to consider my condition an improvement over those that preceded it? Would I be grateful to those who went before me?'

III. IMPLICATIONS FOR HEALTH POLICY

All these considerations appear quite abstract but they may have relevance to some of the urgent health care issues under discussion today. In general, a cross-generational perspective adds an important dimension to our thinking about the right to health care. Elsewhere I have argued for this right and for the responsibility of a developed society like our own to guarantee equal access to high quality health care services. Rational and impartial persons, I believe, would naturally seek to protect themselves against serious ill health. To do this they would insist upon the provision of high quality health services and they would require that access to these services not be made a function of the unequal income shares they might tolerate on other, less pressing grounds ([14, 25, 34]).

Like other arguments for a right to health care this one is essentially present-oriented, focusing on our responsibility to contemporaries. But what happens to these arguments and to the right itself when they are set in the context of an intergenerational moral perspective? Does the longer-term view substantiate this right or call it into question? The answer, I think, is somewhat equivocal. While a cross-generational perspective supports a continuing effort to expand people's access to good health care services, it also calls into question some of the priorities in existing or projected health care systems.

On the positive side, freedom from disease and the effective functioning we often associate with health certainly count among the values we want to promote in the future. Although there are important disagreements about the causes for the improvement of health in recent history, most persons see this improvement as one of the brightest aspects of our legacy from the past. Moreover, since access to high quality health care and personal medical services is still largely based on income and is not yet available to all, it seems reasonable to view equal access to these services as the next major item on our social agenda. It follows that our obligation to the future involves extending and deepening a basic right to health care for all.

Similar thinking underlies efforts to sketch a just pattern of health care development over time. One proposed pattern, for example, has a society in the stage of underdevelopment devote its resources to efficient programs of mass health care including improved nutrition, sanitation, immunization and family planning. As resources increase, more emphasis is placed on personal medical services and high technology research and therapy. Additional growth in income and productivity simply extrapolates this pattern into the future ([5], [14], pp. 119 f.). Not surprisingly, projections of this sort tend to support our own health care priorities by reading them into the indefinite future.

This kind of simple extrapolation has much to say for it, reflecting our rational concern with good quality personal medical care. Since there will probably always be a need for such services it is natural to envision health care improvement in these familiar terms. Nevertheless, one wonders whether this kind of view is really sensitive to the long term. After all, an intergenerationally responsible perspective requires more than reading our wishes into the future. It involves looking at ourselves from that vantage point and asking whether our present conduct would be acceptable to us if we were our own descendants. Applying this test, simple extrapolations of our present health care priorities raise some sharp questions.

IV. THE COST OF MEDICAL CARE

The most obvious difficulty with any health care program that merely extrapolates from existing approaches is the matter of cost. In our own society we are beginning to learn the lesson that an open-ended commitment to quality medical care for all can be enormously expensive. So long as budgets are limited, quality and equality of care may be divergent goals. Of course, the fact that personal health care and acute services are expensive is not itself an argument against them. If impartial persons are presumed willing to affirm a right to personal health care in the present, it is partly because they value this kind of protection more than they do the additional disposable income that might otherwise come their way. And this may also be true as they make choices across time.

Nevertheless, a cross-generational perspective adds some additional reasons for caution in our commitments. First, because where highly technological, labor-intensive medicine is concerned, the decision to do virtually everything we become able to do can easily lead us to relinquish nearly all future discretionary income. Our recent experience with the hemodialysis program indicates that even one technologically demanding program can reach mammoth proportions within the space of a single generation [18]. Not long ago concern with a similar growth of a commitment led to second thoughts about a program for the development and deployment of a totally implantable artificial heart ([1], pp. 212 f.). Although the fact that programs like this can become costly does not mean they are wrong, it suggests that major commitments to extensive curative or palliative efforts must be scrutinized for their sustainability into the future as well as their compatibility with other future commitments we would like to make toward health.

A cross-generational perspective also requires us to regard our present expenditures and commitments with a clear sense of our economic responsibility to the future. I have already mentioned our duty to assist in leaving our descendants better off than ourselves. In discussions of cross-generational ethics, this requirement has usually supported a responsibility to put aside savings for the future (in the form of investment, capital goods and the like). Thus, Rawls argues for a 'just savings' principle requiring each generation to set aside a fair share of its product for the future *before* establishing present income shares and the 'social minimum' of welfare and transfer payments ([32], pp. 284–293). This obligation to the future suggests that the permissible level of health related expenditures cannot be established through a reasonable division of all existing resources as though contemporary or near-term needs were all that had to be considered. Rather, allocation must take into account a fair deduction for savings. If this seems to subordinate the needs of sick persons in the present to crass income considerations, we should remember that economic resources are the basis of future health care services ([6], p. 1628). While a cross-generational perspective does not rule out programs of intensive medical care, therefore, it adds an additional constraint to our contemporary decision-making.

V. EUGENIC CONSIDERATIONS

This kind of limitation on expenditure in the name of the future is not unique to the health care area. To some extent a long-term view calls into question most kinds of present consumption that do not clearly benefit the future. However, a second cautionary implication resulting from a cross-generational perspective is somewhat unique to health care. This is the fact that many present health care efforts, especially intensive therapeutic or palliative services, pose medical risks to the future. Paradoxically, by striving to prevent disease or to provide care for those who are ill today, we may increase the incidence of disease in the future.

This paradox partly arises from the familiar facts of genetic science and the theory of (negative) eugenics. Since health care, especially curative or palliative services, makes it possible for individuals who would otherwise succumb to genetically-caused diseases to survive and reproduce, it also renders it more likely that future persons will be susceptible to similar disease conditions [21]. The emphasis here is on probability, not certainty. For one thing, we do not yet know enough about many features of genetic inheritance confidently to predict their expression. Nor are all inherited traits that produce disease unequivocally bad even in health terms, as the familiar case of sickle cell disease, which is associated with greater resistance to malaria, evidences.

Finally, we have reason to hope that medical science in the future will make progress in treating these diseases at their source, through forms of genetic engineering. Nevertheless, all these qualifications do not entitle us to discount the risk. It is reasonable (and I think responsible) to suppose that genetically-induced diseases will remain a problem well into the future and that present intensive medical services will help perpetuate dysgenic traits. If this is so, our very efforts to relieve suffering for some persons today run the risk of thrusting that suffering on others in the future.

But it is not just in the familar area of eugenic theory that present medical care efforts contribute to future illness. Many complex modern therapies, precisely because of their invasiveness, threaten the health of persons remote in time from those being treated. One familiar example is the post-war use of the drug DES to stabilize pregnancies, with the consequence that years later many daughters born from these pregnancies have begun to experience serious internal disorders. Similar risks are now recognized in the use of radio- and chemotherapy to treat malignancies, while another less familiar illustration of the problem is the use of plutonium as a power source for cardiac pacemakers and the proposed artificial heart. This latter program was halted in its developmental stage because of serious technical difficulties. But among the additional problems noted by critics was the fact that use of plutonium would not only create a risk of cancer or genetic damage for the recipient (who might willingly accept this to extend his life), but also for those healthy persons with whom he came in contact ([1], pp. 121, 132). Indeed, the genetic damage caused by widespread use of devices like this could have ill effects well into the future.

Recent discussions of medical care have given considerable attention to the problem of iatrogenic or physician-induced illness. As far as I know, however, these discussions have usually focused on the penalty treatment imposes on the patient receiving it. But the examples just mentioned indicate that treatment can also threaten persons besides the patient. This suggests the need for a new concept of 'external' or 'third party' iatrogenesis to describe the way in which powerful therapies can inflict harm on 'medically innocent' bystanders, some of whom may be born years after the treatment is employed.

Because this problem of 'external iatrogenesis' is largely the result of intensive curative or palliative medical services, a cross-generational perspective renders these morally suspect. On a strictly utilitarian view, this combination of dysgenic and iatrogenic effects might even furnish a decisive argument against the provision of these medical services. After all, if several generations' neglect of the medically disadvantaged (for example, those prone to hemophilia, youthful diabetes or childhood malignancies) and a forebearance from all externally iatrogenic therapies could improve the health of many future generations, would not this effect the greatest good? As we have seen, however, a utilitarian approach is no more acceptable for intergenerational matters than it is for use within a single generation. While rational impartial persons, who must conceive of themselves as existing anywhere in time, would naturally be concerned to protect themselves from genetic disease in the future, they would also be unwilling to imperil themselves should they be victims of serious diseases in the present for which therapy is available. And they would probably be very reluctant to neglect the immediate health care needs of children ([14], pp. 121 f.).

These future-oriented arguments against the provision of therapy are not, therefore, decisive. They merely represent one additional counter-indication to health care programs that rely heavily on curative services of this sort. Recently, there has been a good deal of criticism of our society's heavy involvement with illness care or 'rescue' medicine. Some of this criticism has been voiced by so-called 'therapeutic nihilists' who view all such services as inherently harmful [20]. But other — and I think more responsible — critics have lamented our relative inattention to the preventive aspects of health care, especially environmental medicine, occupational safety and health education. Sometimes the need for greater emphasis on prevention has been defended on economic grounds, sometimes with reference to historical data indicating that the most substantial gains to health in modern times were made through broad environmental, immunological and nutritional efforts [26]. If a cross-generational perspective is invoked here it provides a further reason for favoring preventive measures that protect essentially healthy populations over curative services, or at least over those therapies with significant dysgenic or iatrogenic effects.

In suggesting this priority, I cannot strongly enough emphasize the fact that future-oriented considerations are at best one small part of the complex health care priority decisions we may be called upon to make. Given our rational self-concern under conditions of impartiality, we would be foolish to advocate the medical abandonment of whole classes of persons on the basis of distant and necessarily conjectural harm. But if we are making very finely balanced social decisions where we are likely to face serious present consequences whichever way we choose, then future-oriented considerations may help further direct our thinking. For example, in deciding between a vital health-education program with demonstrable life saving potential and an extensive effort to sustain infants with defective immune systems, the fact that the latter has long term dysgenic effects is of some importance. And the same may be true if we are choosing between a program to eliminate a known mutagen from the environment or another to develop a marginally superior chemotherapy. If the same number of cancer deaths in the present might be averted in either case, the long term benefits count *for* the environmental program and the possible iatrogenic effects *against* the chemotherapy.

VI. ADDITIONAL PRIORITIES

An intergenerationally responsible perspective, therefore, represents not just a new burden thrust upon us in the making of difficult allocative decisions in health care. It can also assist us in making those decisions by identifying further considerations to enrich our thinking. Indeed, I believe that by trying to take into account the long term implications of our choices, we can readily identify at least several other priorities to add to our present thinking about health care allocation. One is a priority on environmental medicine (and the epidemiological research needed to back it up); a second is on basic biomedical research; and a third is on health care programs of special benefit to lower income groups. These do not exhaust the implications of a long term view, but they stand out as deserving special emphasis.

1. Environmental Medicine

The argument for a priority on environmental medicine (and the forms of research needed to support it) is buttressed by several future-regarding considerations. For one thing, promoting a safe environment seems at least free of the harmful effects for future generations we have seen to be associated with other health care efforts. More positively, a safe environment is itself a major contribution to the health of the future. Along with the industrial development that has been a key factor in economic progress has come a proliferation of new chemical and radiological threats to health. Asbestos, polychlorinated biphenyls, vinyl chloride, methylcholanthrene and benzypyrene are some of the now familiar mutagens in our environment. Overarching these specific pollutants are the problems of general harm to our protective atmosphere and the world-wide increase in ionizing radiation associated with modern industrial processes. I have already indicated our moral obligation to leave our descendants an improved quality of life. Now, while economic progress is a partial fulfillment of this responsibility, it would be far less desirable if it left in its wake serious environmentally induced health problems. Programs to minimize or eliminate environmental hazards make good sense even in the very short run ([27], p. 173; [35], pp. 269 f.), but an intergenerational view further enhances their priority.

2. Biomedical Research

This same obligation to the progressive improvement of the welfare of our descendants supports a priority on basic biomedical research. Part of the reason for this priority is the place that knowledge has in our legacy to the future. In many ways, present generations impose virtually irreversible evils on the future merely by living and consuming in a finite environment. Our substantial consumption of cheap fossil fuels is a familiar illustration of this problem. Nevertheless, by contributing to the fund of human knowledge – in this case, for example, by developing new energy technologies – we can still leave those who follow us in improved and improving circumstances. Apart from adequate food, clean air and water, then, knowledge is about the most important thing we can leave our descendants ([24], p. 40), and this would seem to be particularly true in the area of health care. Progress in health has been one of the most – perhaps the most – felicitous aspects of recent human history. If this progress is to continue, basic biomedical research is probably essential. Though there is much to do in less research-intensive areas (health education, for example), many of the residual problems in developed societies, including congenital disorders, genetic diseases and problems of aging ([26], pp. 21–28), are particularly demanding in research terms. More than ever, therefore, fulfilling an obligation to improved health will require a substantial commitment to research.

A further reason supporting this commitment refers back to our economic obligations to the future. We already noted that much contemporary health care, with its focus on the treatment or palliation of existing disease conditions, is enormously expensive. Improving both the health status and economic circumstances of our descendants, therefore, requires replacing these 'halfway technologies', as Lewis Thomas has called them ([41], pp. 31–36), with technologies capable of stopping disease before it starts or before it significantly impairs functioning. The select areas where medical research has accomplished this – the development of polio vaccine or tuberculosis chemotherapy are examples – remind us that biomedical research cannot only help improve the health of our descendants but can also enable them to spend less of their productivity on health care itself.

In stressing research, I am once again indicating only a *prima facie* consideration when we make complex allocative decisions. A cross-generational perspective does not give research absolute priority even over inefficient forms of medical care or palliation. But where choices must be made, it can give research an edge over embarking on costly new therapeutic ventures. (Of course, we perhaps should keep in mind here the ironic truth that a social commitment to costly therapeutic or maintenance programs may sometimes be *politically* the best way of stimulating support for the research needed to eliminate these programs in the first place.) In addition, this perspective can encourage us to respect the vast research establishment represented in this country by organizations like the National Institutes of Health. As one writer has noted, this and similar establishments elsewhere are a 'world resource' that we must not allow to deteriorate through under-financing or neglect ([10], p. 233). An intergenerationally responsible view which sees this kind of establishment and the knowledge it produces as a major part of our legacy to the future furnishes another reason for resisting efforts to direct resources from basic research to other seemingly more beneficial immediate health care programs.

3. The Health Care Needs of Lower Income Individuals
Finally, I would argue, a future-oriented moral approach provides independent grounds for giving priority to the health care needs of lower income groups in the present. Quite apart from an intergenerational perspective, there are good moral reasons for emphasizing the health care needs of the economically less-advantaged. A moral approach which requires a self-interested but impartial assessment of social policies leads to a special concern with individuals of this sort because under conditions of impartiality it seems rational to try to protect oneself from the serious evils associated with poverty rather than to reach for marginal gain from a more advantaged position. In Rawls's contract theory, this reasoning supports a maximin (maximum minimum) approach to questions of social distribution, with policies measured by how well they serve the long term interests of lower income individuals. Departures from a direct preference for less-advantaged groups are justified only if they serve to make those in these groups eventually better off economically than they would otherwise be ([32], pp. 150–161).

In the area of health policy, this same cautious, self-protective reasoning seems to favor the health care needs of lower income individuals. True, there is a serious question of just who the 'least-off' are where health care is concerned. Are they the economically disadvantaged or are they those (whatever their income level) who are most in need of health services? To some, crucial differences in priority have appeared to rest on the answer to this question ([7, 39]). As I have pointed out elsewhere, however, for most practical purposes these two groups are the same: the poor are generally the most in need of medical services ([16], pp. 10 ff.). In our stage of civilization, at least, poverty and ill health exist in a tight interrelationship with poverty causing ill health and ill health helping to perpetuate poverty [24]. Although the question of who should be considered 'least-off' for health policy purposes still requires theoretical resolution, therefore, we will probably not err from a rationally defensible priority by giving special attention to the significant health care needs of lower income groups. In our society this suggests an emphasis on problems of hypertension, prenatal medicine and pediatric care.

One reason for stressing the health care needs of lower income individuals, therefore, is that this is a basic priority of social justice. But at least two other considerations drawn from an understanding of intergenerational justice further support this priority. First, there is the matter that intergenerational moral concern is a relatively new and demanding aspect of our total moral responsibility. As such it is very likely to be neglected or resisted. Partly because we have not thought much about these responsibilities in the past, and also partly because we have never been in a position before to do so much harm to the future, we are not accustomed to exercizing our intergenerational conscience. The debates over energy and environmental policy demonstrate how extremely difficult it is to ask individuals or societies involved in bitter struggle over limited resources to restrain their demands in the name of future persons. But since lower income individuals are usually locked in the most desperate struggle merely to maintain themselves, it is often especially difficult to elicit their cooperation on behalf of long-term goals. The resistance of lower income workers to job-threatening environmental programs is an illustration of this problem. In the area of health policy, lower income individuals may similarly resist programs oriented toward the future if they perceive these to be neglectful of their own health care needs. Fortunately, many aspects of an intergenerationally responsible health policy, including the emphasis on environmental medicine, also tend to benefit lower income individuals. But this is not likely always to be the case, and we cannot expect less advantaged persons to be enthusiastic about commitments to basic research or restrained in their support of expensive palliative programs if they see acute care efforts as their best chance to benefit from social expenditures on health. To generate cooperation with intergenerational justice, in other words, requires some concern with social justice in the present.

A final major reason for emphasizing the health care needs of lower income groups is the fact that one of our key, non-medical responsibilities to the future is the control of human population growth. We have already seen the basis for this in our most fundamental reasoning about obligations to the future. In this context, however, what is important is the fact that lower income groups both in this country and abroad have the highest rates of fertility. Usually, this association is explained in terms of the cultural backwardness of these groups and the economic pressures that prompt couples to have large families. But it has also been pointed out that the absence of good quality personal medical care contributes as well to high fertility. Although mass preventive measures (improved nutrition, sanitation, immunization) have reduced aggregate mortality rates, on a family-by-family basis the fact that individual children can still be lost because of inadequate medical care prompts couples to continue older reproductive patterns, over-insuring for the sake of family survival [40].

One part of the solution to the intergenerationally disturbing problem of rapid population growth, therefore, lies in placing an emphasis on the provision of adequate personal medical care to lower income families. When we also consider that medical care facilities providing a broad range of basic services are also widely regarded as the best context for the provision of family planning services, we have a further reason for setting this priority. Whether in a national perspective, or as part of our nation's foreign aid responsibility to underdeveloped countries, a focus on the medical needs of the less-advantaged may be one of the most effective means of fulfilling our responsibility to the future.

VII. CONCLUSION

The question of the extent of our obligations to future generations represents a relatively new area of moral reflection. Also relatively new is the question of how a society responding more and more to the conviction that there is a basic human right to health care should allocate its necessarily finite health care resources. In the preceding remarks I have tried to suggest that efforts to think about these two different questions can actually complement one another. Stressing our essential duty to strive for an improved quality of life for our descendants, including a state of better health, I have argued that this duty has important implications for our contemporary health care priority decisions. On the negative side, a future-oriented perspective tends to militate against programs involving an emphasis on costly curative or palliative services. More positively, this perspective supports programs stressing prevention, environmental medicine, basic biomedical research and the provision of fundamental medical services to lower income individuals. All these are only *prima facie* considerations, however, and they do not rule out a substantial degree of emphasis on acute medical care in a developed society like our own. The anti-utilitarian thrust of an impartial reasoning process applied across generations tends to discourage the abandonment of existing individuals in need for the sake of preponderant future benefit, and this acts as a check upon the economic and eugenic considerations that otherwise counsel against a heavy emphasis on acute medical care in the present.

An intergenerationally responsible perspective does not therefore promise a dramatic reversal of our present health care priorities. Rather, it appears to support a moderate shift toward objectives that have been widely advocated on other grounds as well. Of course, given the magnitude of many major social decisions on health care, even a moderate shift can be very significant in terms of the resources involved. Because this is so, how we reflect upon our obligations to the future can have a profound impact on the shape of our evolving health care system. These remarks, and the priorities they have been held to support, are meant primarily as a stimulus to that ongoing process of reflection.

Dartmouth College
Hanover, New Hampshire

BIBLIOGRAPHY

1. Artificial Heart Assessment Panel: 1973, *The Report of the Artificial Heart Assessment Panel of the National Heart and Lung Institute*, Department of Health, Education and Welfare, Washington, D. C., Publication No. (NIH) 74–191.
2. Baier, K.: 1965, *The Moral Point of View*, abridged edition, Random House, New York.
3. Bell, D.: 1978, *The Cultural Contradictions of Capitalism*, Basic Books, New York.
4. Branson, R.: 1978, 'Theories of Justice and Health Care', *Encyclopedia of Bioethics* 2, 630–637.
5. Bryant, J. H.: 1977, 'Principles of Justice as a Basis for Conceptualizing a Health Care System', *International Journal of Health Services* 7, 707–719.
6. Card, W. I. and Mooney, G. H.: 1977, 'What is the Monetary Value of a Human Life', *British Medical Journal* 2, 1627–1629.
7. Daniels, N.: 1979, 'Rights to Health Care and Distributive Justice: Programmatic Worries', *Journal of Medicine and Philosophy* 4 (June), 174–191.
8. Delattere, E.: 1972, 'Rights, Responsibilities and Future Persons', *Ethics* 82, 254–258.

9. Feinberg, J.: 1974, 'The Rights of Animals and Unborn Generations', in W. T. Blackstone (ed.), *Philosophy and the Environmental Crisis*, University of Georgia Press, Athens, Georgia, pp. 43–68.
10. Frederickson, D. S.: 1977, 'Health and the Search for New Knowledge', in *Priorities for the Use of Resources in Medicine*, Department of Health, Education and Welfare, Washington, D.C., pp. 229–235.
11. Govier, T.: 1979, 'What Should We Do About Future People?' *American Philosophical Quarterly* 16, 105–113.
12. Golding, M.: 1972, 'Obligations to Future Generations', *Monist* 56, 85–99.
13. Golding, M.: 1979, 'Obligations to Future Generations', *Encyclopedia of Bioethics* 2, 507–512.
14. Green, R. M.: 1976, 'Health Care and Justice in Contract Theory Perspective', in R. M. Veatch and R. Branson (eds.), *Ethics and Health Policy*, Ballinger Publishing Co., Cambridge, Mass, pp. 111–126.
15. Green, R. M.: 1977, 'Intergenerational Distributive Justice and Environmental Responsibility', *Bioscience* 27, 260–265.
16. Green, R. M.: 1977, 'The Nuclear-Powered Totally Implantable Artificial Heart – A Rawlsian Reassessment', Unpublished paper for the Conference on Policy Making in Health Resource Allocation: Concepts, Values, Methods, Sponsored by the National Legal Center for Bioethics, Washington, D. C.
17. Green, R. M.: 1975, *Population Growth and Justice*, Scholars Press, Missoula, Montana.
18. Hastings Center: 1976, 'Long-Term Dialysis Programs: New Selection Criteria, New Problems', *Hastings Center Report* 6, 8–13.
19. Hubin, D. C.: 1976, 'Justice and Future Generations., *Philosophy and Public Affairs* 6, 70–83.
20. Illich, I.: 1976, *Medical Nemesis*, Pantheon Books, New York.
21. Imaizumi, Y. *et al.*: 1970, 'Variability and Heritability of Human Fertility', *Annals of Human Genetics* 33, 251–259.
22. Jones, H.: 1976, 'Genetic Endowment and Obligations to Future Generations', *Social Theory and Practice* 4, 29–46.
23. Kavka, G.: 1978, 'The Futurity Problem', in R. E. Sikora and B. Barry (eds.), *Obligations to Future Generations*, Temple University Press, Philadelphia, pp. 186–203.
24. Kitigawa, E. M. and Hauser, P. M.: 1973, *Differential Mortality in the United States: A Study in Socioeconomic Epidemiology*, Harvard University Press, Cambridge, Mass.
25. McCreadie, C.: 1976, 'Rawlsian Justice and the Financing of the National Health Service', *Journal of Social Policy* 5, 113–131.
26. McKeown, T.: 1976, *The Role of Medicine: Dream, Mirage or Nemesis?*, Nuffield Provincial Hospitals Trust, London.
27. Mechanic, D.: 1979, *Future Issues in Health Care*, The Free Press, New York.
28. Mill, J. S.: 1864, *Utilitarianism*, Longman, London.
29. Narveson, J.; 1978, 'Future People and Us', in R. E. Sikora and B. Barry (eds.), *Obligations to Future Generations*, Temple University Press, Philadelphia, pp. 38–60.
30. Narveson, J.: 1973, 'Moral Problems of Population', *Monist* 57, 62–86.

31. Passmore, J.: 1974, *Man's Responsibility for Nature*, Pantheon Books, New York.
32. Rawls, J.: 1971, *A Theory of Justice*, The Belknap Press of Harvard University Press, Cambridge, Mass.
33. Routley, R. and Routley, V.: 1978, 'Nuclear Energy and Obligations to the Future', *Inquiry* 21, 166–173.
34. Shelton, R. L.: 1978, 'Human Rights and Distributive Justice in Health Care Delivery', *Journal of Medical Ethics* 4, 165–171.
35. Sidel, V. W. and Sidel, R.: 1977, *A Healthy State*, Pantheon Books, New York.
36. Sidgwick, H.: 1907, *The Methods of Ethics*, 7th ed., Dover Press, New York.
37. Sperber, J. P.: 1977, 'Obligations to Future Generations: Explorations and Problemata', *Journal of Value Inquiry* 11, 104–116.
38. Stearns, J. B.: 1972, 'Ecology and the Indefinite Unborn', *Monist* 56, 612–625.
39. Stewart, M. M.: 1977, 'Problems in Applying Principles of Justice to Health Care Systems', *International Journal of Health Services* 7, 727–731.
40. Taylor, H. and Berelson, B.: 1971, 'Comprehensive Family Planning Services Based on Maternal/Child Health Services: A Feasibility Study for a World Program', *Studies in Family Planning* 2, 21–54.
41. Thomas, L.: 1974, *The Lives of a Cell: Notes of a Biology Watcher*, The Viking Press, New York.
42. Vetter, H.: 1969, 'The Production of Children as a Problem of Utilitarian Ethics', *Inquiry* 12, 445–447.
43. Warnock, J.: 1971, *The Object of Morality*, Methuen, London.

EARL E. SHELP

JUSTICE: A MORAL TEST FOR HEALTH CARE AND
HEALTH POLICY

I. INTRODUCTION

The distribution of health care services in the United States has come under increasing critical review during the past decade. Even though most observers would admit that the North American system of health care is imperfect, only a few would argue that the perceived defects are severe enough to justify tossing it on the garbage dump of meritless institutions and customs. A recognition of the excellence of certain practices has counseled a cautious and prudential response to contemporary calls for total reorganization. The language of reason and reform, rather than the rhetoric of revolution, has received a more sympathetic hearing by providers of health care services and makers of public policy.

The protests and appeals for reform advanced by critics of the present system have focused on its equity and accountability. They cite six general areas of concern. First, they charge that essential services are not geographically accessible to all citizens. Second, they identify the high and rising costs of services as a barrier to needed care for a significant number of people. Third, they suggest that the mix of available services (generalists and specialists) is unresponsive to real personal and social health needs. Fourth, they question the efficacy of the system to improve health levels as determined by traditional statistical measures (e.g., life expectancy and infant mortality). Fifth, they assert that the services are not of a uniform quality due to ineffective peer review, dependence on fee-for-service, and consumer ignorance. Sixth, they maintain that the disparate character of services provided to the welfare recipient as compared to the private patient is unconscionable. Critics claim on the basis of these allegations and others that the good of health care services is being denied to individuals or classes of individuals on morally irrelevant grounds. In sum, they charge that the present system has certain characteristics of injustice that ought to be remedied.

These professions of injustice may spring from an inherent intuition of fairness or from a seasoned reflection on the nature and order of human association. Whichever is the case, commentators at best have only pointed to a description of a just health care system by drawing attention to its opposites. This may be all that is reasonable to expect or possible to achieve. And its contribution, however incomplete, deserves appreciation. In the final analysis, John Stuart Mill may have been correct that the distinguishing character of justice is impossible to discern and that it "is best defined by its opposites" ([22], p. 43). The difficulty of the task in this case does not, however, discharge the urgency of its undertaking. The character of a society and its health care institutions will be significantly influenced by the form and content of these deliberations.

This essay is rather limited in scope. I shall not attempt to resolve the numerous questions about theories of justice or the criteria for distributing the health care benefits and burdens of society. Neither shall I propose a novel approach to justice which successfully avoids the difficulties of other theories and incorporates their advantages. Finally, I shall not evaluate the positive and negative attributes of the several suggested morally relevant criteria for distribution which supposedly satisfies the requirements of justice. I shall, however, attempt to develop the thesis that the principle of justice is a moral test for the institutions of health care and health policy. Other non-moral considerations, such as supply, efficiency, and cost benefits, also may be important to this decision. And other moral tests, such as the principle of beneficence and the virtue of compassion, may be important to a reasoned moral judgment. Justice and these additional moral and non-moral factors would appear in creative tension, each effecting modifications or compromises in the other, but when considered together contribute to a description and judgment of the character of North American health care and health policy.

In order to develop my thesis, I shall do three things. First, I shall identify a notion of the intrinsic value of human beings and the means instrumental to human welfare as the subject of distributive justice which together constitute common components essential to four theories of justice. I shall cite selected representatives of each theory who express these common notions. The described theories may have other similarities and obviously many differences but these will be considered irrelevant to the present task. Second, I shall sketch an understanding of a moral judgment, health care, and health policy. Finally, I shall discuss the adequacy of justice as a moral standard and consider its applicability as a test for health care.

II. THEORIES OF JUSTICE

A thorough presentation of selected major theories of justice is far beyond the scope of this paper and, in certain ways, irrelevant to its task.[1] A general and simplified statement of each is sufficient to demonstrate two points at which they share a basic concern. No attempt will be made to assess the merit of each theory or to elaborate the particular interpretations or statements which exist within each theoretical framework. Rather, these descriptions are believed to accurately state the distinctive thrust of each school without unnecessarily complicating them with digressions on the nuances of each. The four ethical traditions to be reviewed are the utilitarian, contractarian, egalitarian, and Judeo-Christian.

Utilitarian

Utilitarian theories can be broadly defined as those ethical theories which hold that the principle of utility is the ultimate standard of morality. The desired end of human action, according to utilitarians, is the greatest balance of non-moral good (however conceived) over evil ([17], pp. 34–35). Thus, utilitarian justice requires that social benefits and burdens ought to be distributed in order to effect this end.[2]

John Stuart Mill was perhaps the most influential advocate of this method of moral decision-making. In his classic work *Utilitarianism*, Mill understood justice as embracing the notion of one's personal right or security. This right was understood in terms of a valid claim against others and constituted the essential differences between justice and beneficence which admitted no such claim. He thought that the idea of justice supposed two things: (1) a rule of conduct intended for the good of human beings, and (2) a natural, human sentiment of retaliation for harms to oneself or to others, including all persons, as a result of an extended sympathy and intelligent self-interest ([22], pp. 47–49).

Mill recognized that individual competing claims of one's right could be based on merit and effort. Merit was understood to define what one should receive. Effort was understood to define what society should give. He argued that social utility alone was capable of negotiating these competing claims since any choice between the two on the grounds of justice (one's right) would be purely arbitrary. Thus, he accepted justice as a rule of conduct which respected one's right but which was ultimately grounded on utility as its chief, most sacred, and binding part ([22], pp. 53–54).

Mill based the priority of justice on its object of human well-being. He considered justice to include "certain classes of moral rules which concern the *essentials of human well-being* more nearly, and are therefore of more absolute obligation, than any other rules for the guidance of life, and the notion which we have found to be of the essence of the idea of justice – that of a right residing in an individual – implies and testifies to this more binding obligation" ([22], p. 54; emphasis added). Apart from his ultimate loyalty to utility, it thus appears that Mill viewed the allocation of the means essential to human well-being (happiness) as a proper subject of justice on the basis of the value attributed to a quantitative and qualitative human existence. Whatever goods and services are understood as essential to human welfare would, according to this interpretation, be subject to a distribution which would maximize the net good for the society as a whole.

Joseph Fletcher [14] and, more recently, Tom Beauchamp and Ruth Faden [3] have demonstrated how the questions of the macroallocation of health care can be addressed from the perspective of utilitarian theory. The statement of Beauchamp and Faden, which adopts a 'cost/benefit analysis constrained by a decent minimum criterion', is the superior of the two exercises and more worthy of careful consideration.

Contractarian

The second major theory of justice to be reviewed is that of John Rawls who built on the traditional theory of social contract represented by Locke, Rousseau, and Kant ([25], p. 11). In his influential work, modestly titled with the indefinite article *A Theory of Justice*, Rawls characterized justice as "the first virtue of social institutions" and the principles of social justice as "a way of assigning rights and duties in the basic institutions of society and ... [*which also*] define the appropriate distribution of the benefits and burdens of social cooperation" ([25], pp. 3–4). He concluded after an extended review that two distributional principles would be adopted by a society if the parties were 'rational and mutually disinterested', in an 'original position' of equal status, and 'behind a veil of ignorance' ([25], pp. 12–13). These two principles are ultimately formulated as: (1) "Each person is to have an equal right to the most extensive total system of equal basic liberties compatible with a similar system of liberty for all," and (2) "Social and economic inequalities are to be arranged so that they are both: (a) to the greatest benefit of the least advantaged, consistent with the just savings principle, and (b) attached to offices and positions open to all under conditions of fair equality of opportunity" ([25], p. 302).

His 'ideal theory' would require that all "social primary goods – liberty and opportunity, income and wealth, and the bases of self-respect – are to be distributed equally unless an unequal distribution of any or all of these goods is to the advantage of the least favored" ([25], p. 303). One should note that Rawls does not identify health care as a 'social primary good' but rather as a 'natural primary good' not subject to the distributional principles.

Rawls's theory is in opposition to the utilitarian bias which esteems the aggregate good even if it is at the expense of the few. Justice, according to Rawls, is uncompromising and will admit injustices only when necessary to avoid greater injustices. His theory affirms the inviolability of each person which cannot even be overridden by the welfare of the whole society ([25], pp. 3–4). His principles are applicable to social institutions whose just operations would provide for the greatest individual minimal share of the 'social primary goods' which serve as a basis for interpersonal comparisons of well-being and a means by which the ends of individual rational plans are realized ([25], pp. 152–157; 90–95). It therefore appears that Rawls believed his principles would mediate conflicting individual interests and assist through institutional structures in the protection and promotion of individual welfare.

Ronald Green has applied the contract theory of Rawls to health care, claiming that health care can and ought to be considered a primary social good subject to the distributive principles. Green challenged the view that health is a function of natural contingencies and thus outside of the scope of social distribution. He argued, instead, that in modern society health is in many ways a function of medical technology and social decisions. As such, health care has an instrumental value to the fulfillment of one's life plan and is thus subject to a distribution according to the Rawlsian principles of justice [18].[3]

Egalitarian

Egalitarian, like utilitarian, theories of justice have taken several forms.[4] What each shares is a formal principle of an equality of persons. The formal principle alone is not taken as an adequate definition of distributive justice since it does not identify specifically how individuals are equal, who belongs in a certain category of equality, or how specifically individuals are to be treated within a category. The validity of the formal premise also is questionable on empirical grounds. Few people would claim that all persons share, among other things, an equal natural endowment, ability, intelligence, interest, or industry. Perhaps it would be disadvantageous to human welfare if they did. Thus, egalitarians tend to interpret their formal principle of equality as an affirmation of an equal intrinsic *worth* of each human being. Granting this, egalitarians proceed to supplement the formal principle with morally relevant qualifiers which, alone or in some combination, are believed to constitute a basis upon which comparative distributions of social benefits and burdens may be justified, e.g., need and merit.

William Frankena considered equality, understood as equal consideration, equal opportunity, and equality before the law, as built into the concept of justice. The end of justice for the individual is the good life as individually conceived. He argued that the positive interest of a just society is to provide conditions within which its members can have an equal chance to achieve to the extent capable their own version of the good life ([16], p. 14). Equal concern for the good life of each member of society may require different treatment for each since Frankena does not equate equal treatment with identical treatment ([15], p. 20). If an equality of worth in the moral community is granted, then, egalitarians suggest, each individual should have an equal opportunity for well-being and freedom ([32], p. 49). Thus, a distribution of the burdens and benefits of society, including health care if it is viewed as a component of or contributor to the good life as individually defined, would be just to the extent that it met the requirements of these principles.

Applications of egalitarianism to health care have been made by Gene Outka and Robert Veatch. Outka concluded that justice in health care requires similar treatment for similar cases [23]. Veatch's egalitarian formulation of "justice requires health care sufficient to provide an opportunity for a level of health equal, as far as possible, to the health of others . . ." ([30], p. 136). Integral to each view are the corollary notions of equal intrinsic human worth and a concern to assure an equal opportunity to the means necessary to preserve health.

Judeo-Christian

The Western religious traditions of Judaism and Christianity have participated in the discussion of social justice out of a conviction that God wills justice in the social order. The human community is to be characterized by righteousness modeled after the example of God. Among the standards of righteousness is a concern for justice as a distinctive of human relationships. The eighth century prophet Isaiah called upon Israel to "cease to do evil, learn to do good; seek justice, correct oppression; defend the fatherless, plead for the widow" (*Isaiah* 1: 16—17). The prophet Micah was even more explicit in identifying the command of God. "He has showed you, O man, what is good; and what does the Lord require of you but to do justice, . . . " (*Mic.* 6: 8). The Gospel of Matthew records Jesus as sharing these prophetic concerns. He conceived of justice and righteousness as God-like characters which ought to find expression in daily life (*Matthew* 5–6).

The Hebrew word for justice is *mishpat* and the word for righteousness is *tsedeq*. The terms are different but their meanings are closely connected. They are often used in the Hebrew Scriptures in parallel constructions, one supplementing and defining the other. The Greek equivalent of *tsedeq* is *dikaiousune* which is translated as righteousness or justice depending on the context. What is important to this discussion is that these are dynamic terms of relationship which portray God's conduct in covenant with man and the character that God requires of human association.

Central to the theologies of Judaism and Christianity is an assertion of the value of each individual. The creation narratives in *Genesis* 1–2 declare the 'goodness' (worth) of human life which merits protection regardless of one's past conduct or present social standing (*Genesis* 4: 8–15). Jesus gave witness to this fundamental tenet as he ministered to those individuals considered outside the boundaries of social and religious respectability. This aspect of his ministry is particularly evident in the Gospel of Luke. The scope of God's love is viewed as universal which seeks the positive good of each even if one is presently outside of the exemplary community of faith (*Romans* 12: 20; I *Timothy* 2: 1–4; I *Peter* 2: 12–16). All are to be treated on the basis of the value they possess in the sight of God rather than the sight of man (*Matthew* 8: 1–4; 20: 1–16). Thus God is understood as willing the welfare of all in recognition of the intrinsic worth of each.

Justice, according to this tradition, is an expression of love and acts out the claims of righteousness. Justice requires a recognition of the dignity of each person who ought to be afforded an equal opportunity to realize God's will for well-being which requires indiscriminate treatment (*Leviticus* 19: 15; *Matthew* 5: 43–45). Actions and policies which divide the benefits of the community so as to promote individual human welfare are considered just.

Theological ethicists have commented on justice and the distribution of health care. Much of my own work has been in this area ([27, 28]). Charles Curran has provided a similar analysis from a Roman Catholic perspective [11]. And Robert Shelton has argued for a moral right to health care that government ought to guarantee [29].

Shared Judgments

Each of the four constructions (utilitarian, contractarian, egalitarian, Judeo-Christian) have as essential components a two-fold concern. The first is to affirm the inherent worth of human beings. The second is to identify the subject of distributive justice as the benefits of society which are perceived as instrumental to human welfare. This is not to suggest that the theories agree on the relative priority to be given to individual welfare or to the aggregate good. Neither is it to suggest that they are in agreement about the relevant properties upon which to differentiate distributions nor the respective goods subject to the distributive principles. They do, however, share a confidence that justice is an instrument to and protector of autonomy.

Each approach senses that some form of justice is a requirement for social peace and human welfare. The requirements of justice do not stem solely from a favorable disposition which would cause one to advocate another's due. Rather, they accrue from a perception of the dependent and interdependent nature of human existence. One's sense of responsibility to help secure the claims of another springs from a perception, however grounded, of the other's worth and a reciprocity of need and desire for fulfillment that deserves acknowledgement. This intuition of human dignity and purpose provides these theories of justice with a distinctively moral aspect. It constitutes a basis of a moral demand for just consideration in the distribution of social benefits perceived as instrumental to human welfare.

Values and Justice

Theories of justice are value-laden. Whatever scope is given to justice and which principles of distribution are accepted will reflect what has been determined as valuable. Justice is conceived, a human invention, not a human discovery or, necessarily, a revelation of absolutes from God (this of course depends on one's doctrine of revelation). The truth of this assertion is demonstrated by the fact of several notions of justice rather than only one. Each reflects a different view of the universe, an ideal of the good life, and the means necessary to its approximation. A perfect theory of justice has not been constructed, perhaps it never will.

C. Perelman's analysis of justice is instructive in this regard. He considered any system of justice as no more than the development of one or more values which satisfy a rational need for coherence and regularity. Justice is a rational virtue, the product of reason in action. It establishes what has value, prescribes a standard, and appeals to values other than the value of justice for justification ([24], pp. 52–60). If this account is correct, then the dynamic character of justice is partially explained. As values within a society change, theories of justice are influenced. As the means considered instrumental to human wellbeing are identified and accepted the scope of justice will be expanded or contracted to govern their just distribution. The instrumental goods and services considered subject to the principles of distributive justice will be selected in response to dominant social values and on the basis of their relative contribution to human welfare, however defined.

This process of selection and allocation may involve tragic choices as described by Guido Calabresi and Philip Bobbitt ([9], pp. 17–28; 195–198). Pluralistic societies embrace many values which may on occasion come into conflict, especially in situations of scarce resources. In such instances, the society attempts to preserve in its choices (policies) the appearance that its fundamental values by which it defines itself are intact and have not been abandoned.

This subterfuge occurs in two stages. The first-order determination sets how much of a scarce good will be produced because of absolute natural scarcity or on the basis of relative priorities within a large framework of ultimate natural scarcities. The second-order determination decides who shall get what is available. If, as was the case with the early use of kidney machines in the United States, the first-order determination restricts the supply of a good, then the postulate that a particular human life is priceless is contradicted. Further, some distributional ideals of the society (for example, equality of opportunity as opposed to ability to pay) are violated unless the actual distribution is widely supported and accepted as justified.

This decision-making process, when successful, gives the appearance that an allocation is necessary, unavoidable, and unfortunate, but not a choice. The conflict of fundamental values is obscured and the allocation is not viewed as tragic. Yet, when the subterfuge is exposed the tragic dilemma reappears since fundamental values are seen in conflict once again. New choices are made in an effort to remove the moral contradiction within the present allocation. This method generates hope that the expected loss from the original allocation will not be realized, that necessity can be ultimately evaded, and that values currently degraded will not be abandoned.

Tragic choices may involve those goods and services that are considered related to a society's fundamental values in an essential way. Life, health, and well-being (ignoring the substantial problems of definition) receive high priority within American society. Many like to think that life is without price, that health can be purchased, and that well-being is the *summum bonum*. It is not too difficult to understand how medical care is becoming increasingly associated with these ends as necessary to their realization. The more a link can be established between medical care and life, health, and well-being, the more medical care will appear subject to social allocation. More will be said about this later. It is sufficient to note for the present that an appreciation of this perceived relationship and an appreciation of the viable, value-laden, and evolutionary scope and nature of justice may help to explain recent efforts to include medical care as a target of distributive justice.

Summary

The first objective for this essay has now been met. It has been shown that each of the examined four theories of justice can agree on at least two essential points. The first is that each shares a premise of individual human worth and a vision of human welfare. The second is that each has as the subject of distributive principles of justice the benefits of society considered instrumental to the envisioned human welfare. I suggested, in agreement with Perelman, that these notions of justice are value-laden, viable, and evolutionary. They respond to changed values and material circumstance which may help to account for an increased number of 'welfare rights,' including medical care, grounded in theories of justice. The second section of this essay will examine the relationship of justice to health care and health policy.

III. MORAL TESTS

The adequacy of distributive justice as a moral test for medical care and health policy will be now discussed. The suitability of the standard or standards of justice to perform this service has been assumed in much of the literature and implied in the present discussion. The deliberations in this section will seek to establish a justification for subjecting health care distributional practices and health policy to considerations of justice. Also, comment will be made on the sufficiency of justice as a measure of morality.

Moral Judgments

If one is to issue a moral test of certain rules or practices, then one must adopt a moral point of view. Kurt Baier held that one adopts a moral point of view "when one considers one's social order as grounded by a set of supreme guiding principles for all its members, and thinks of these members as rational agents willing to be guided by such principles as long as they have reasons for doing so as good as have any of the other members" ([1], pp. 5, 46). This description may not be totally satisfactory but it does help to isolate the nature of moral judgments. A moral judgment is distinguishable from other sorts of judgments by the kind of reasons offered in its support. A moral judgment of health care would not appeal, for example, to the standards of efficiency or legal entitlement as a basis of evaluation or defense. Rather, an appeal would be made to moral principles or rules that have some force within a given society.

James Childress cautioned against an equal weighting of each of the four criteria proposed by Baier and others to establish a moral rule or principle (action-guide). Childress, basicallly in interaction with William Frankena, considered the formal conditions of prescriptivity, universality, and the social material condition of other-regardingness as necessary and sufficient for moral judgments and moral rules and principles. He warned against, however, an acceptance of the 'overriding' condition except in a weak sense. He opposed a strong or absolute sense of overridingness for two reasons. First, it limits, by definition, the debate about which sorts of consideration should be given the most weight. Second, it limits again by definition, a full discussion of the reasons why one ought to abide by the rule or principle ([10], pp. 7, 5–13). This qualification appears of particular importance to a consideration of justice and health care. What some would suggest as required of health care by justice may not, in fact, be of overriding importance within the larger universe of moral concerns. An answer to the questions of how much and why will establish, according to this view, the priority or overridingness of justice against other considerations regarding health care and other goods such as efficiency, opportunity costs, or beneficence.

The social material condition of other-regardingness serves to narrow the range of rules and principles which count as moral. This condition requires a consideration of other persons and a concern for their welfare. It underscores the common notion that moral evaluation is primarily interested in the treatment of human beings and the effect that actions, policies and conditions of life have upon human welfare. David Little and Sumner Twiss viewed 'welfare' as the 'seed stock' of morality and only allowed it a rather restricted core of meaning, limited to the "provision of 'material' conditions necessary to maintain life" ([21], p. 55). This orientation of morality to basic concerns of human welfare corresponds to an intuition that this is and ought to be a distinguishing focus of moral evaluation.

Before specifically considering justice as a moral test for health care and health policies, a few preliminary remarks about the relationship of these pursuits to welfare or well-being are in order.

Health Care and Health Policy
The phrases 'health care' and 'health policy' have been used throughout these remarks without specifying their meaning. 'Justice' and 'distributive justice'

have been used synonymously. Little would be gained for the purpose of this essay from an excursus which differentiated 'health care' from 'medical care', 'justice' from 'distributive justice', or which delineated 'health policy'. However, a few comments about 'health care' and 'health policy' appear productive to the successful completion of this investigation.

There is a popular association, and in some cases an equation, of health care and health. The prevailing assumption is that one seeks and obtains the former in order to attain the latter. The practice, in some cases, has been that when someone falls ill no price is considered too great to pay for therapy which offers some prospect for cure. No distances are considered too great to travel or no intervention is seen as optional since health, and maybe life, is at stake. Allegiance to the idea that life is priceless has conferred a similar position of prestige on health care. The wealthy and the poor, the lettered and the unlettered, the city dweller and the rural dweller, alike issue claims for health care when the stakes are seen as so high. Whether or not health care is actually as important to health or well-being as this sketch implies remains to be proven.

The assumption that health care can always produce health, and is therefore of great value, is threatened. A belief in the efficacy of health care alone to produce health has been challenged by knowledge demonstrating the impact of personal lifestyles,[5] genetic endowment, and environmental hazards to health status and mortality. Ivan Illich even claimed that, in fact, there is an inverse relationship of health to health care [20]. All of this raises questions about responsibility for health and illness. Further, questions are posed about the worth of health care to health and to what extent health care should be valued over and above other goods which may contribute to well-being.

Similar questions can be asked with regard to health policy, understood, for the present purposes, as the regulations and practices of government designed to preserve and promote the health of the citizens. These policies can be in the form of mandatory vaccination, environmental controls, substance regulation, building codes, occupational safety requirements, health care financing programs, and many others. Of importance to this presentation is the implicit statement within these policies that health and health care are valued by the government and the citizens. Governmental policies do not grant health and health care ultimate priority. This is evidenced by its contradictory tobacco program. On the one hand, the Department of Health, Education, and Welfare requires cigarette manufacturers to publicize a warning that cigarette smoking has been determined hazardous to one's health. On the other hand, the Department of Agriculture subsidizes the farmer who grows tobacco. Health is esteemed in public policy but not necessarily at the expense of all other endeavors. This is amply illustrated by the almost annual effort to legislate a national health insurance which falls to defeat in the ostensible names of economy, efficiency, and free enterprise. On the one hand, health care and health are valued. On the other hand, some other good is valued more highly.

This brief discussion suggests two general observations. The first is that health is valued in our society and that health care is perceived, reasonably or not, as a means to health. The second is that health care, broadly understood and distinguished from health, is valued as a social good and perceived as potentially subject to some expanded form of governmental allocation. These statements are subject to a great many qualifications which would probably not be objectionable. The intent here is not to analyse and evaluate the arguments that could be formulated about these observations. Rather, the intent is simply to assert the *perception* of health care as a valued good.

Justice, Health Care, and Health Policy
The above summary discussion of justice suggested that each of the four statements share a twofold interest. Each school presumed a greater or lesser inherent worth of human beings which serves as a grounding and justification for ethical theory, moral rules, and moral principles. Each presentation also reflected an interest in the means considered instrumental to human welfare or well-being as the proper subject of distributive justice. Whether one is a utilitarian, contractarian, egalitarian, or religious ethicist, broad general agreement can be reached in at least these two areas. The company will be parted when the implications of these common concerns are sketched out and relative priorities are assigned. Nevertheless, they speak in concert to a concern for the right (just) allocation of social benefits which respects human dignity and potential by awarding each one's due insofar as possible.

The discussion further suggested that conceptions of justice are value statements which reflect the *Weltanschauung* of the theorist. The goods and services valued as instrumental to human welfare are incorporated into the orbit of justice as they are identified and receive broad approval. The boundaries of justice thus expand and contract in response to values that gain prominence or drop in prestige. The goods and services associated with these values are then included or excluded as subjects of distributive principles of justice.

The aspect of moral principles considered especially important to the use of justice as a basis of moral judgment for health care was other-regardingness.

The other-regarding condition focuses a moral judgment on the effects that actions, policies, and conditions of life have on human welfare. The more that certain material conditions of life are considered essential, rather than instrumental, to human welfare the more their allocation will be thought of as subject to judgments based on justice. Since health care is valued and perceived as an instrumental, or in some cases essential, component of or means to human welfare, attempts have been made to ground its distribution in principles of justice even to the point of claiming its receipt as a right.

A task of theories of distributive justice is to test the basis and methods of allocation from a moral point of view ([26], p. 7). Some theorists, like Rawls, would consider justice as the acid test of social institutions. The basic structure of society and its major social institutions would be judged morally according to how effectively arbitrary distinctions were removed as bases of distribution and how effectively its rules properly balance competing claims to social benefits ([25], pp. 3, 5). Yet, to limit morality and moral evaluation, which Rawls does not, to justice would be impoverishing.

The propriety of justice as a test of health care and health policy is affirmed. The perception of the good or value of each institution to human welfare warrants the measure. Justice can reasonably be held as a virtue of society and its institutions, including health care and health policy, but it ought not be their sole virtue or measure of morality. There is more to morality than justice, particularly with regard to human pain and suffering which may elicit more than one's duty to meet another's due. To limit the moral test of health care and health policy to justice runs the risk of limiting morality to duty at the expense of other desirable attributes such as beneficence, sacrifice, and compassion which share an interest in human welfare and which may compete with justice for applicability and priority. The dispute about the nature of the duties of justice as absolute or prima facie, perfect or imperfect points to the insufficiency and difficulty of applying this standard alone as the sole test of morality.

Justice is concerned with the proper use of power or advantage with regard to the treatment of the powerless or disadvantaged. Justice requires that the defenseless not be exploited or denied arbitrarily the benefits of social cooperation out of respect for their inherent human worth. To this extent, health care and health policy must be just as a minimal standard for allocation. A society and its institutions can be just without being ideal, but a society and its institutions cannot be ideal without being just ([15], pp. 1–3). Health care and health policy can be just without being morally excellent, but health care and health policy cannot be morally excellent without being just. Justice is, indeed, a moral test among others of society and social institutions. It is both a goal or ideal which inspires its approximation and a norm which requires its establishment.

IV. CONCLUSION

Health care and health policy are proper subjects of justice. Regardless to which of the selected theories examined one subscribes or the particular interpretation of it preferred, there can be general agreement that institutions which contribute to human welfare can be subject to a moral test of justice. A considered judgment based on the standard of justice may or may not be sufficient to establish the morality of a society, an institution and its practices. Yet, such institutions, and their practices, vital to human welfare cannot escape a review of justice.

Institute of Religion and
  Baylor College of Medicine
Texas Medical Center
Houston, Texas

NOTES

[1] More complete, but still brief, surveys are presented in the articles by A. Buchanan and F. Carney in this volume. See also [8, 13].
[2] For a critique of utilitarian justice see [26].
[3] For a critical examination of Rawls see [2], and of Rawls and Green see [12].
[4] For example, see [4, 7, 15, 16, and 32].
[5] See [5] and [6] for a more complete statement. See [31] for a discussion of the ethical issues.

BIBLIOGRAPHY

1. Baier, K.: 1978, 'Ethical Principles and Their Validity', in *The Belmont Report*, Appendix Vol. 1, National Commission for the Protection of Human Subjects of Biomedical and Behavioral Research, DHEW Publication No. (OS) 78–0013, Washington, D.C., pp. 5, 1–55.
2. Barry, B.: 1973, *The Liberal Theory of Justice*, Clarendon Press, Oxford.
3. Beauchamp, T. L., and Faden, R. R.: 1979, 'The Right to Health and the Right to Health Care', *The Journal of Medicine and Philosophy* 4 (June), 118–131.
4. Bedau, H. A.: 1971, 'Radical Egalitarianism', in H. A. Bedau (ed.), *Justice and Equality*, Prentice-Hall, Inc., Englewood Cliffs, pp. 168–180.
5. Belloc, N. B.: 1973, 'Relationship of Health Practices and Mortality', *Preventive Medicine* 2, 67–81.

6. Belloc, N. B. and Breslow, L.: 1972, 'Relationship of Physical Health Status and Health Practices', *Preventive Medicine* **1**, 409–421.
7. Benn, S. I.: 1971, 'Egalitarianism and the Equal Consideration of Interests', in H. A. Bedau (ed.), *Justice and Equality*, Prentice-Hall, Inc., Englewood Cliffs, pp. 152–167.
8. Branson, R.: 1978, 'Theories of Justice and Health Care', in W. Reich (ed.), *The Encyclopedia of Bioethics*, Vol. 2, Macmillan Publishing Co., and the Free Press, New York, pp. 630–637.
9. Calabresi, G. and Bobbitt, P.: 1978, *Tragic Choices*, W. W. Norton, N.Y.
10. Childress, J. F.: 1978, 'The Identification of Ethical Principles', *The Belmont Report*, Appendix Vol. 1, National Commission for the Protection of Human Subjects of Biomedical and Behavioral Research, DHEW Publication No. (OS) 78–0013, Washington, D.C., pp. 7, 1–35.
11. Curran, C. E.: 1979, 'The Right to Health Care and Distributive Justice', in *Transition and Tradition in Moral Theology*, University of Notre Dame Press, Notre Dame, pp. 139–167.
12. Daniels, D.: 1979, 'Rights to Health Care and Distributive Justice: Programmatic Worries', *The Journal of Medicine and Philosophy* **4** (June), 174–191.
13. Feinberg, J.: 1978, 'Justice', in W. Reich (ed.), *The Encyclopedia of Bioethics*, Vol. 2, Macmillan Publishing Co., and The Free Press, N.Y., pp. 802–811.
14. Fletcher, J.: 1976, 'Ethics and Health Care Delivery: Computers and Distributive Justice', in R. M. Veatch and R. Branson (eds.), *Ethics and Health Policy*, Ballinger Publishing Co., Cambridge, pp. 99–109.
15. Frankena, W. K.: 1962, 'The Concept of Social Justice', in R. B. Brandt (ed.), *Social Justice*, Prentice-Hall, Inc., Englewood Cliffs, pp. 1–29.
16. Frankena, W. K.: 1966, 'Some Beliefs About Justice', The Lindley Lecture, Department of Philosophy, University of Kansas, Lawrence.
17. Frankena, W. K.: 1973, *Ethics*, 2nd ed., Prentice-Hall, Inc., Englewood Cliffs, N.J.
18. Green, R. M.: 1976, 'Health Care and Justice in Contract Theory Perspective', in R. M. Veatch and R. Branson (eds.), *Ethics and Health Policy*, Ballinger Publishing Co., Cambridge, pp. 111–126.
19. Hume, D.: 1975, *Enquiries Concerning Human Understanding and Concerning The Principles of Morals*, Third Edition, L. A. Selby-Bigge (ed.), Clarendon Press, Oxford.
20. Illich, I.: 1976, *Medical Nemesis*, Pantheon Books, N.Y.
21. Little, D. and Twiss, S. B.: 1973, 'Basic Terms in the Study of Religious Ethics', in G. Outka and J. P. Reeder, Jr., (eds.), *Religion and Morality*, Anchor Books, Garden City, N.Y., pp. 35–77.
22. Mill, J. S.: 1971, *Utilitarianism*, Samuel Gorovitz (ed.), Bobbs-Merrill Co., Inc., Indianapolis.
23. Outka, G.: 1974, 'Social Justice and Equal Access to Health Care', *Journal of Religious Ethics* **2** (Spring), 11–32.
24. Perelman, C.: 1963, *The Idea of Justice and the Problem of Argument*, The Humanities Press, N.Y.
25. Rawls, J.: 1971, *A Theory of Justice*, Belknap Press of Harvard University Press, Cambridge.

26. Rescher, N.: 1966, *Distributive Justice: A Constructive Critique of the Utilitarian Theory of Distribution*, Bobbs-Merrill Co., Inc., Indianapolis.
27. Shelp, E. E.: 1976, 'An Inquiry Into Christian Ethical Sanctions for the "Right to Health Care"', Unpublished Ph.D. Dissertation, Southern Baptist Theological Seminary, Louisville.
28. Shelp, E. E.: 1978, 'A Biblical Commentary on Justice in Health Care', *Catalyst Cassette Tapes* **10** (June).
29. Shelton, R. L.: 1978, 'Human Rights and Distributive Justice in Health Care Delivery', *Journal of Medical Ethics* **4**, 165–171.
30. Veatch, R. M.: 1976, 'What Is a "Just" Health Care Delivery?' in R. M. Veatch and R. Branson (eds.), *Ethics and Health Policy*, Ballinger Publishing Co., Cambridge, pp. 127–153.
31. Veatch, R. M.: 1980, 'Voluntary Risks to Health: The Ethical Issues', *Journal of the American Medical Association* **243** (January 4), 50–55.
32. Vlastos, G.: 1962, 'Justice and Equality', in R. B. Brandt (ed.), *Social Justice*, Prentice-Hall, Inc., Englewood Cliffs, pp. 31–72.

NOTES ON CONTRIBUTORS

Michael D. Bayles, Ph.D., is Director, Westminster Institute for Ethics and Human Values, Westminster College, London, Canada, and Professor of Philosophy, The University of Western Ontario, Canada.

Baruch Brody, Ph.D., is Professor of Philosophy, Rice University, Houston, Texas.

Allen E. Buchanan, Ph.D., is Associate Professor of Philosophy, University of Minnesota, Minneapolis, Minnesota.

Fredrick S. Carney, Ph.D., is Professor of Christian Ethics, Perkins School of Theology, Southern Methodist University, Dallas, Texas.

Kim Carney, Ph.D., is Professor of Economics at the University of Texas at Arlington, Texas, and Adjunct Professor at Southwestern Medical School, Dallas, Texas.

Eric J. Cassell, M.D., F.A.C.P., is Clinical Professor at Cornell University Medical College, New York, New York.

James F. Childress, Ph.D., is Professor of Religious Studies, University of Virginia, Charlottesville, Virginia.

H. Tristram Engelhardt, Jr., Ph.D., M.D., is Rosemary Kennedy Professor of the Philosophy of Medicine, Kennedy Institute of Ethics, Georgetown University, Washington, D.C.

Martin P. Golding, Ph.D., is Professor of Philosophy, Duke University, Durham, North Carolina.

Ronald M. Green, Ph.D., is Associate Professor, Department of Religion, Dartmouth College, and Adjunct Assistant Professor, Department of Community Medicine, Dartmouth Medical School, Hanover, New Jersey.

Albert R. Jonsen, Ph.D., is Professor of Ethics in Medicine, Health Policy Program, School of Medicine, University of California, San Francisco, California.

Marc Lappé, Ph.D., is Director, Hazard Alert Systems, Department of Health Services, Berkeley, California.

Karen Lebacqz, Ph.D., is Associate Professor of Christian Ethics, Pacific School of Religion, Berkeley, California.

Laurence B. McCullough, Ph.D., is Associate Director, Division of Health and Humanities, Department of Community and Family Medicine, School of

Medicine, and Senior Research Scholar, Kennedy Institute of Ethics, Center for Bioethics, Georgetown University, Washington, D.C.

Earl E. Shelp, Ph.D., is Assistant Professor of Bioethics, Institute of Religion and Baylor College of Medicine at the Texas Medical Center, Houston, Texas.

INDEX

abortion 66, 87, 122, 181–182
Abraham 26–27, 39–40
agent morality 144
Al-Ghazali 42
American Medical Association 64, 67
amniocentesis 90–91
Aquinas, Thomas 30–31, 43
Aristotle 25, 30, 44, 112
Aristotelianism 43
Augustine of Hippo 42, 48–49
autonomous agent 9
autonomy 10, 179, 220
Averroes 43
Avicenna 42

Baier, Kurt 222–223
Bard, Samuel 61, 63–65, 67
Basil of Caesarea 46
Bayles, Michael D. xii, 109–117
Beauchamp, Tom 54–55, 216
Becker, Lawrence 144
beneficence-benevolence xiv, 55, 63–64, 109, 111, 114, 116, 130, 179, 215, 223, 226
Bentham, Jeremy 76
Bice, T. W. 173
bioethics ix, 3–4, 49, 67
Blackstone, William 32
Bobbitt, Philip 221
Bracton, H. 29
Branson, Roy ix, 67
Breslow, Lester 147
Brody, Baruch xiii, 151–159
Brown, Louise 89
Buchanan, Allen E. x, 3–21
Buddhism 37

Calabresi, Guido 221
California Department of Health Services xii, 86

Calvin, John 49
Campbell, A. G. M. 95, 100
Captain Cook 123
Carney, Frederick S. x, 37–50
Carney, Kim xiv, 161–178
Cassell, Eric J. xi, 72–82
Celsus 27–28
Certificate-of-Need 172–174, 176
Chadwick, Edwin 76
Chamberlain, Wilt 13
charity x, 14, 19, 45, 56, 57, 76, 129, 153
Childress, James F. xiii, 55, 65, 139–150, 223
Christianity 37–50, 218–219
Cicero 25, 27–30, 48
Clement of Alexandria 42
Cohen, Hermann 43
compassion xi, 75, 79–82, 226
conservatism 17
Consumer Price Index 162–163
cosmology 39–40
cost-benefit xii, 54, 62, 106, 145, 216
cost containment xiv, 163–164, 171, 174, 177
covenant 45, 55, 62, 67, 219
Curran, Charles 219

Daube, David 26
defective newborns xii, 3, 182
    medical indications policy 96, 102
    quality of life 95–96, 98–105
    rights 96
    value of life 95–96
deontology 5, 6, 10
Department of Health Services 86
DES 92, 202
difference principle 6, 18–19
dignitas 28–29
dignity 25, 220, 225

dikaiousune 219
Duff, Raymond 95, 100
Dworkin, Ronald 5, 23, 29, 33
dying xii–xiii
　allocation of resources 112–115
　definition 114
　euthanasia 111
　rights of 110–112, 116–117
　suicide 110–111
　truth-telling 111–112

Economic Stabilization Program 171–172
efficiency xii, 109, 140, 142, 145, 156, 223, 225
effort 131, 180, 215
Eli Lilly and Company 92
Engelhardt, H. Tristram, Jr. xi, xiii, xv, 53, 121–137
Enlightenment 61, 65, 68
entitlements x, xiv, 30, 57, 127–129, 134, 156, 168, 223
envy 131
equal access xiii, 130–132
equality xii, 7, 15, 48, 78–79, 81, 98–99, 101–102, 104, 112, 143, 145–146, 151, 180–181, 197, 217
ethics 123–124

Faden, Ruth 54–55, 216
fairness x, xiii, 6–10, 122, 129, 145
fair opportunity 6, 18–20
fate 77, 79
fertility xii, 89–90, 92
first-order 139, 221
Fletcher, Joseph ix, 67, 216
Flew, Anthony 140
Food and Consumer Services 85
Frank, Johann Peter 58–60, 66–68
Frankena, William 102, 218, 223
French Revolution xi, 56–57
Fried, Charles 129, 141, 143–144, 147

Gabriol, Ibn 42
goal-rational 55, 143
Golding, Martin P. x, 23–25

Green, Ronald M. xiv, 67, 129, 193–211, 217
Gregory, James 61–65, 68
Gregory, John 61
Gregory of Nyssa 42
Gregory the Great 45
Gross National Product 161–162
Grotius, Hugo 32

health 75–76, 140, 198, 224
health care
　access 163–164, 167–169, 174, 176–177
　allocation xiii–xiv, 76, 81, 128–132, 144, 199–201, 209
　competition xiv, 171, 174–176
　costs xiv, 53, 75, 142, 151, 161–163, 200–201, 205, 213
　eugenics 201–202, 209
　financing 168–170, 175–177
　iatrogenic 93, 202–203
　indigent xiii, 153–154, 157–158, 206–208, 213
　macro-allocation x, xiii–xiv, 3, 18, 124, 134, 139–148, 157, 216
　micro-allocation x, 3, 18, 17, 139
　minimum xiii, 62, 121, 130–131, 141, 216
　regulation xiv, 170–174, 176
　religious incentives 38, 45, 47
health care systems xiii, 121–123
　criticism 14, 52–53, 213
　egalitarian 122–123, 130–132
　free market 121, 128–129
　mixed 121, 129–130
health maintenance organization 171, 174–176
Health Planning and Resources Development Act of 1974 172
health status 166–167, 224
Health Systems Agency 172–174
Hellegers, André 67
HEW Secretary's Task Force on the Compensation of Injured Research Subjects 180, 189–190
Hinduism 37
Hobbes, Thomas 33, 196

INDEX

homunculus 88
human research xiv
    disadvantaged 183–185
    equal treatment 181–182, 186–187
    fetus 181–182
    injury 188–190
    IRB 190
    merit 182–183
    prisoners 185–188
    racial factors 187
    remuneration 186–187
    selection of subjects 179–180
    socio-economic factors 187
human well-being xii, 215–222, 225
Hume, David 61–62
Hutcheson, Francis 61, 63

Illich, Ivan 224
infant mortality 84–86, 168, 213
informed consent xiv, 110, 179, 189
Ingelfinger, Franz 146
institutional medical ethics xi, 51–52, 56, 63–64, 66, 68
in vitro fertilization 89–90
Islam 37–50
ius 24, 27–32
Jesus Christ 41, 83
Jonas, Hans 67, 179, 184
Jonsen, Albert R. xii, 67, 95–107
Judaism 37–50, 218–219
justice
    compensatory vii, 115–116, 145, 181
    definition 24–26, 30, 40, 77, 80, 98, 146, 165, 179, 214
    distributive ix–xii, 25, 38–39, 44, 54, 76, 86, 89–92, 112–115, 145–146, 151–153, 181, 214, 220–221
    egalitarian 217–218
    fairness x, 6–10, 25, 216–217
    intergenerational xiv, 193, 198, 204–209
    particular 25
    principles xiv–xv, 5–14, 43, 180, 216, 218
    retributive xi–xii, 54, 86, 92–93, 181
    social xii, 86, 207, 216
    theological x, 37–50, 217–218 (see theological justice)
    theories x, xiv–xv, 3–4, 20, 214–215, 225
    universal 25
    utilitarian 4–6, 93–94, 215–216

Kant, Immanuel 9–10, 27, 43, 49, 109–110, 216
Kantorowicz, Hermann 28
Kass, Leon 147
Kennedy Foundation 95

Lappé, Marc xi, 83–94
Lazarus, Moritz 43
Lebacqz, Karen xiv, 179–191
Leibniz, G. W. von 58
Lesius, Leonhardus 30
Lesky, Erna 58
liberty xiii, 6, 18, 48, 59, 146–148
libertarianism x, 10–14, 16–17, 155–156
    quasi xiv, 156–157
licensing laws 17
life expectancy 168, 213
Little, David 223
Locke, John 11, 57, 156, 216
Lockean Proviso 11, 16
love xi, 45–46, 79–80, 219

MacIntyre, Alasdair 123
McCall, N. 162
McCormick, Richard 95–96
McCullough, Laurence B. xi, 51–71
Magna Carta 30
Maimonides, Moses 43, 49
Marxism 37
May, William F. 62
maximin 8, 206
medical ethics 51–52
medical police xi, 57–60, 65–67
Medicaid 121, 151, 158, 167–169, 177
Medicare 121, 151, 158, 167–169
medicine
    biomedical research xv, 205–206, 209

crisis xiii, 16, 75, 142, 165, 203
environmental xv, 204–205, 209
preventive xiii, 16, 75, 124, 141–148, 203–204, 209
mentally infirm 185
mercy xi, 45–46, 75, 79–82
merit x, 25, 28–30, 31, 180–183, 215, 218
Mill, John Stuart 5, 23, 154, 214–216
minimal state 12–13
mishpat 26, 219
Mohammed 41, 47
moral judgments xv, 214, 222–223, 225–226
Moses 40–41

National Commission for the Protection of Human Subjects of Biomedical and Behavioral Research 180–188, 190
National Foundation/March of Dimes 85
national health insurance 75–76, 147
National Health Service 76
National Institutes of Health 206
natural lottery xiii, 125–127, 133–134
natural resources 156–157
need xiv, 14, 38, 101–102, 114, 124–125, 135, 141, 151–154, 158, 165–166, 180–181, 183–185, 195, 206–208, 213, 218
neonatal intensive care xii, 3, 95, 102, 105–106
neonatology 95, 104–105
neo-Platonism 42
Niebuhr, H. Richard 48
Niebuhr, Reinhold 47
Nozick, Robert 10–14, 16, 17, 124, 126, 129
nursing care 112–113
Nuremberg Code 179

obligation xi, xiii–xiv, 39, 43–44, 55, 62, 68, 115, 193–197, 215–216
Office of Economic Opportunity 167
Origen 42
original position 7–10, 129
Outka, Gene 146, 165, 218

paternalism 80, 147
patient-physician relationship 51, 55, 60–68, 75–82, 139
Paulus 27
Percival, Thomas 51
Perelman, C. 220, 222
personhood 78–79, 81, 184
Philo of Alexandria 42
Placentinus 28–29
Plato 24, 42, 48, 124
Plattner, Marc 152, 157
population growth 207–208
prenatal life xi–xii
    attitudes toward 83–85
    development 88–89
    maternal relationship 86, 90, 92–93
    medical care 15, 86, 89, 93
    rights 87, 93
    status xii, 87
productivity 180
Professional Standards Review Organizations 174
protective agencies 12
public health policy xi–xii, xv, 39, 65, 68, 78, 139, 142, 164, 206–207, 213, 222, 224–226

quality of life xii, xv, 84, 198, 204, 208

Ramsey, Paul ix, 96–97, 99, 102, 140, 184
Rau, Wolfgang Thomas 58–60
Rawls, John x, xiv, 6–10, 13, 17–20, 54, 66–67, 78, 126–127, 129, 195–196, 201, 206, 216–217, 226
rectification 12
Reich, Warren, 95–96
respect for persons xiv, 179
right x, 24, 27, 30, 33
righteousness 26, 40, 218–219
rights x, 5, 10, 14, 23, 25–33, 109–110, 215
    claim 14, 30–31, 109, 111–112
    fetal 87, 93
    health care x–xi, xiii–xiv, 14–20,

53–57, 67, 122–123, 128, 132–134, 153–155, 157, 193, 199, 208
liberty 30–31, 33, 109, 111
natural 30–31, 57
newborn 85
objective 24, 29, 33
patients 62–63, 109
powers 31–32
property xiii–xiv, 6, 11–14, 30, 152, 155–157
subjective 24, 27–29, 31–33
to die xii, 110–111
utilitarian 5–6, 14–16, 53–54, 154
risk-benefit xiv, 180, 183, 187
Ritschl, Albrecht 43
Roe v. Wade xii, 87
Roman Catholic Church 83
Rosen, George 58
Rousseau, Jean Jacques 60, 216

Saadya 42, 49
Saint Paul 41
Salkever, D. S. 173
Sarah 26–27
Schaffner, Kenneth 79
Schleiermacher, Friedrich 43
Schultz, Fritz 27
Scitovsky, A. 162
Seagram Distilleries 92
Second Isaiah 47
second-order 139, 221
secular humanism 37
Shelp, Earl E. ix–xvi, 213–229
Shelton, Robert 219
side constraint 123–124, 134–135
Siegler, Mark 55, 62, 65, 67–68
Simonides 24
Smith, Adam 12, 169
social contract 8
social primary goods 7, 8, 10, 18–19, 129, 217
Stason, W. B. 145
state of nature 12
statistical lives 142–143
Stoics 24
Suarez, Francisco 27, 31, 32

Supreme Court 23, 87, 88
symbolic value 142–144

taxpayers xiii, 151–152
teleology 5, 10, 43
theological justice
  Aristotelianism 43
  content 44–47
  corrective 45
  distributive 38–39, 44–45
  fiducial qualities 38–39, 44
  Greek philosophy 41–44
  love 45–46
  mercy 45–46
  revelation 40–41
  scope 39
  theodicy 49
  transcendence 47–49
Thomas, Lewis 205
Tillich, Paul 46
tragic choices 221
tsedeq 45, 219
Twiss, Sumner 223

Ulpian 24–25, 27–30, 32, 48
unfortunate 125–127, 131, 134
utilitarianism x–xi, 4–6, 13–16, 92–93, 194–195
utility 4–5, 10, 14–16, 55, 68, 104, 114, 142, 154, 180–181, 215–216

value-rational 66, 65, 68, 143–144
values xiii, xv, 39, 46–47, 114, 131, 143, 220–221
Veatch, Robert M. ix, 51, 54–55, 130, 218
veil of ignorance 8–9
Villey, Michel
virtue xi, xv, 13, 25, 29, 39, 43–44, 55, 68, 80, 109, 133, 216, 220, 226
virtuous physician xi, 60–65, 68
Vlastos, Gregory 23

Weber, Max 55, 143
Weinstein, M. D. 145
William of Ockham 32

wisdom xi, 81, 198
Women, Infants, and Children's Supplemental Feeding Program 85
World Health Organization 140, 198
worth 25, 28, 115, 217–219, 225–226

zabat 45
Zachary, R. B. 96
zedekah 26

The Philosophy and Medicine Book Series

Managing Editors

H. Tristram Engelhardt, Jr. and Stuart F. Spicker

1. **Evaluation and Explanation in the Biomedical Sciences**
   1975, vi + 240 pp.   ISBN 90–277–0553–4

2. **Philosophical Dimensions of the Neuro-Medical Sciences**
   1976, vi + 274 pp.   ISBN 90–277–0672–7

3. **Philosophical Medical Ethics: Its Nature and Significance**
   1977, vi + 252 pp.   ISBN 90–277–0772–3

4. **Mental Health: Philosophical Perspectives**
   1978, xxii + 302 pp.   ISBN 90–277–0828–2

5. **Mental Illness: Law and Public Policy**
   1980, xvii + 254 pp.   ISBN 90–277–1057–0

6. **Clinical Judgment: A Critical Appraisal**
   1979, xxvi + 278 pp.   ISBN 90–277–0952–1

7. **Organism, Medicine, and Metaphysics**
   Essays in Honor of Hans Jonas on his 75th Birthday, May 10, 1978
   1978, xxvii + 330 pp.   ISBN 90–277–0823–1

174.2 J98   107021

JUSTICE AND HEALTH CARE